**280**

# Advances in Polymer Science

# Aims and Scope

The series Advances in Polymer Science presents critical reviews of the present and future trends in polymer and biopolymer science. It covers all areas of research in polymer and biopolymer science including chemistry, physical chemistry, physics, and material science.

The thematic volumes are addressed to scientists, whether at universities or in industry, who wish to keep abreast of the important advances in the covered topics.

Advances in Polymer Science enjoys a longstanding tradition and good reputation in its community. Each volume is dedicated to a current topic, and each review critically surveys one aspect of that topic, to place it within the context of the volume. The volumes typically summarize the significant developments of the last 5 to 10 years and discuss them critically, presenting selected examples, explaining and illustrating the important principles, and bringing together many important references of primary literature. On that basis, future research directions in the area can be discussed. Advances in Polymer Science volumes thus are important references for every polymer scientist, as well as for other scientists interested in polymer science - as an introduction to a neighboring field, or as a compilation of detailed information for the specialist.

Review articles for the individual volumes are invited by the volume editors. Single contributions can be specially commissioned.

Readership: Polymer scientists, or scientists in related fields interested in polymer and biopolymer science, at universities or in industry, graduate students.

Special offer:

For all clients with a standing order we offer the electronic form of Advances in Polymer Science free of charge.

More information about this series at http://www.springer.com/series/12

Werner Pauer

Editor

# Polymer Reaction Engineering of Dispersed Systems

Volume I

With contributions by

M. Ahsan Bashir · M. A. Dubé · K.-D. Hungenberg ·
E. Jahns · C. Kiparissides · L. Lei · B.-G. Li ·
T. F. L. McKenna · W. Pauer · Y. Zhang · H. Zhu · S. Zhu

 Springer

*Editor*
Werner Pauer
Institute for Technical and Macromolecular Chemistry
University of Hamburg
Hamburg, Germany

ISSN 0065-3195          ISSN 1436-5030   (electronic)
Advances in Polymer Science
ISBN 978-3-319-73478-1      ISBN 978-3-319-73479-8   (eBook)
https://doi.org/10.1007/978-3-319-73479-8

Library of Congress Control Number: 2018957453

This Springer imprint is published by the registered company Springer Nature Switzerland AG
The registered company address is: Gewerbestrasse 11, 6330 Cham, Switzerland

**Prof. Dr. H.-U. Moritz**

*Prof. Dr. Hans-Ulrich Moritz is a chemist and full professor at the University of Hamburg, and Chair of Technical Macromolecular Chemistry at the University Institute for Technical and Macromolecular Chemistry. He was born in West Berlin on June 30, 1952. His scientific life started in the research group of Prof. Dr. Karl-Heinz Reichert, with a PhD thesis on "Continuous Bead*

*Polymerization of Vinyl Acetate in a Stirred Tubular Reactor," and has since then focused on polymer reaction engineering.*

*Consequently, he defended a postdoctoral thesis on "Computer-Assisted Laboratory Reactor for Emulsion Polymerization of Vinyl Acetate" in 1988.*

*After a short industrial intermezzo, he received his first full professorship in Technical Chemistry and Chemical Process Engineering at the University of Paderborn in 1990. During his scientific career, he published numerous articles and is documented as an inventor in many patents.*

*Besides his scientific life, Prof. Moritz is active as the chairman of the DECHEMA Education Committee for Technical Chemistry (since 2008), the chairman of the DECHEMA Working Committee for Reaction Technology on processes presenting difficulties from a safety perspective (since 2002), an appointed member of the DECHEMA Working Committee on Polyreactions (since 2000), and a member of the Board of Directors and Trustees of the Jung Foundation for Science and Research (since 2011).*

# Preface

Polymer reaction engineering of dispersed systems is a dynamic and broad research field with a high impact on everyday life. This is particularly obvious for dispersion paints but has many more aspects. Consequentially, the present volume of Springer's *Advances in Polymer Science* is intended to provide a review recognizing relevant research in both academic and industry labs.

The occasion to start work on this volume was the upcoming 65th birthday of Prof. Dr. Hans-Ulrich Moritz, to whom the authors dedicate this edition. Prof. Moritz's scientific work covers manifold aspects with a major focus on dispersed systems and reactor design, such as Couette–Taylor reactors and bent tubular reactors for continuous emulsion polymerization. The industrial realization of spray polymerization has relied on the research and development in his labs. Additionally, his interests include special research fields such as chemical safety engineering and online analytics.

José M. Asua describes the current knowledge of emulsion polymerization and moves on to the challenges of the unknown in this multiphase system.

A special chapter on mass transport in complex mixtures is contributed by Klaus Tauer, Chunxiang Wei, Amit Tripathi, and Olga Kiryutina. The mechanism of mass transport is one of the key processes of emulsion polymerization and is currently being controversially discussed. The authors do not just give a literature overview; rather, they dispute the various aspects of mass transport. The field of precipitation polymerization is presented by Liborio Ivano Costa and Giuseppe Storti, with a main emphasis on modeling. They summarize the most relevant aspects of the process, focusing on the free-radical polymerization mechanism, with an emphasis on the key role of radical interphase transport.

Despite an incomplete theoretical description and perhaps a lack of comprehensive understanding, polymerization in dispersed systems is a solid base for economically successful industrial products. There is a high level of empirically elaborated know-how, especially in industry, that may have been condensed into accurate but unpublished theoretical models.

A key factor for success is process monitoring and control. Hence, Eric Frauendorfer, Muhammad Babar, Timo Melchin, and Wolf-Dieter Hergeth summarize and discuss the latest process-monitoring technologies, focusing on their applicability.

Facing the challenges of the unknown, an overview on stimuli-responsive latexes is given by Michael F. Cunningham, Philip G. Jessop, and Ali Darabi. The field of stimuli-responsive latexes opens new application fields with switchable stabilization/destabilization. Therefore, coagulation and stability modeling is presented by Martin Kroupa, Michal Vonka, Miroslav Soos, and Juraj Kosek, as well as by Hua Wu, Dan Wei, and Massimo Morbidelli. Stabilization by Pickering emulsion and development of novel materials therefrom are reviewed by He Zhu, Lei Lei, Bo-Geng Li, and Shiping Zhu.

Shaghayegh Hamzehlou and Jose Ramon Leiza review the field of composite polymer latexes. They elaborate on recent developments in the synthesis and application of composite (hybrid) latex particles, including polymer–polymer and polymer–inorganic latex systems and modeling efforts to simulate the development of particle morphology.

New products with novel characteristics are to be expected as further monomer resources become available, preferably from renewable sources. This emerging research field is reviewed by Yujie Zhang and Marc A. Dubé. They give a comprehensive overview of the technology and show the variety of substance classes from sustainable sources. General aspects of continuous emulsion polymerization are summarized by Werner Pauer, and the field of polyolefin production is discussed by M. Ahsan Bashir and Timothy F. L. McKenna.

PVC is one of the large-scale polymers produced by dispersion polymerization processes. However, in the currently accessible scientific literature, reports on PVC do not nearly have the corresponding volume. Costas Kiparissides thus gives a rare review, gathering knowledge on modeling of the dynamic evolution of the polymerization rate, concomitant average molecular weights, and morphological properties of dispersion PVC.

The contributions are completed by a chapter from Klaus-Dieter Hungenberg and Ekkehard Jahns, who take a look at the future of emulsion polymerization processes from an industrial perspective.

I am confident that this digest on dispersed systems provides a sound compilation of the current state of the art, which can be consulted by both young researchers and experienced practitioners.

I would like to express my very sincere thanks to the authors for their support and timely responses, to the reviewers for their important feedback, and to the whole editorial team of *Advances in Polymer Science* for their valuable assistance.

Special thanks from the editor go to the Hamburg Institute of Technical and Macromolecular Chemistry for its continuous support in completing this volume – in particular, to Prof. Dr. Dr. hc. mult. Walter Kaminsky, Prof. Dr. Patrick Théato, Prof. Dr. G. A. Luinstra, and Mrs. Christina Khenkhar.

Hamburg, Germany                                                                        Werner Pauer
1 June 2017

# Contents

Adv Polym Sci (2018) 280: 1–18
DOI: 10.1007/12_2017_24
© Springer International Publishing AG 2017
Published online: 10 August 2017

# Reactor Concepts for Continuous Emulsion Polymerization

Werner Pauer

**Abstract** Emulsion polymers are products by process. Besides the process and recipe conditions the reactor itself is the main factor that determines the product properties. In today's industrial environment the stirred tank reactor concept still is the main concept in use. This may be due to the easy design of the process and the fact that existing equipment in a plant can be used. But Emulsion Polymerization products are not uniform and reactor design can be taylored to the optimal process/product. This article reviews the wide variety in reactor concepts for continuous emulsion polymerization from the early beginning up to recent developments.

**Keywords** Continous emulsion polymerization • Dispersed systems • Polymerization reactor

## Contents

W. Pauer (✉)
Institute for Technical and Macromolecular Chemistry, University of Hamburg, Bundesstrasse 45, 20146 Hamburg, Germany
e-mail: werner.pauer@uni-hamburg.de

# 1  Introduction

The industrial imitation of a natural rubber latex emulsion in 1912 is commonly accepted as the starting point of heterophase polymerization [1]. At that time, the polymeric structure of polybutadiene had not yet been elucidated [2]. The situation changed in 1927 with the works of Staudinger, who proved the covalent nature of polymeric structures [3].

Since the early beginnings of heterophase polymerization, the progress made in chemical reaction technology is evident. One driving force was knowledge of the influence of temperature on polymer viscosity and, thus, the influence of temperature on product properties [4]. From this, it became obvious that polymers are so-called products by process. Today, it is known that the molecular weight distribution of synthetic heterophase polymers determines the rheological and final product properties. Hence, a key factor in the success story of polymers was the new research field of polymer reaction engineering. Early examples of industrial polymerization concepts were based on batch and semibatch reactors [5, 6]. Soon, continuous reactor concepts for polymerization emerged [7] and then concepts for operation of dispersed phase polymerization in continuous tank reactors [8], continuous tubular reactors [9–11], and continuous tanks in series [12]. An early example of gas phase polymerization of ethylene is documented by Seebold [13]. Thereby, the basic ideas for continuous heterophase polymerization technology were developed and used in continuous heterophase polymerization [14]. A better understanding of what happens in emulsion polymerization was brought by the path-breaking work of Harkins [15–18].

The fundamental basics of kinetics and reactor concepts are included in various encyclopedias [19, 20]. Furthermore, comprehensive books on, or with extended sections on, emulsion polymerization kinetics are available [21–27]. However, research on emulsion polymerization still requires a lot of further activity. Sheibat-Othman gives a critical review on emulsion polymerization modeling [28].

Further reviews are available with focuses on engineering and recipe aspects, as well as theoretical and modeling aspects [29–40]. For continuous miniemulsion and controlled radical polymerization, comprehensive reviews are available that include relevant examples of tubular and continuous stirred tank reactors (CSTRs) [41, 42]. The transfer from a semibatch to continuous process is of special interest, especially for specialty products [30, 43, 44].

Continuous reactor concepts for ethylenic and propylenic polymers are described in a chapter by Bashir and McKenna in this volume [45]. The present chapter gives an overview of reactor concepts for continuous macroemulsion polymerization.

## 2  Stirred Tank and Tank in Series Reactors

Stirred tank reactors are very flexible in terms of use. They can be easy adapted for different operation modes (e.g., batch, semibatch, and continuous) and for special process requirements (e.g., by interconnecting stirred tank reactors), such as dosing needs or residence time distribution needs [46]. Consequently, the first heterophase polymerization reactors were batch and semibatch stirred tank reactors [5, 6]. Driven by military and defense needs, rubber synthesis was pushed by the German and US governments during World War II [47] and resulted in tanks in series of 6–12 tanks [48]. Thus, the first models were successfully realized for this continuous system [49] and later transferred to other monomer systems [50, 51]. The concept of using tanks in series was chosen consistently. On the one hand, heat removal is simplified but, on the other hand, kinetic aspects must be considered. Fikentscher stated that a single stirred tank reactor might be disadvantageous because of back-mixing in continuous emulsion polymerization [52]. Back-mixing causes lower conversion and self-sustained oscillations of the mean particle size, molar mass, and conversion [53].

Furthermore, multiple stationary states are observable in single CSTRs [54]. An overview of experimental work on self-sustained oscillations is given by Omhura et al. [55]. With respect to the works of Harkins, Smith, Ewart, and colleagues the concept of seeded emulsion polymerization was developed to overcome the self-sustained oscillations [16, 18, 56, 57]. Omi showed that a seeder, here a stirred tubular reactor, increases the efficiency of a CSTR [58–60]. Consequently, for CSTR processes, a latex seed is typically dosed into the reactor or a seeding reactor is coupled to the CSTR to avoid oscillations. By this, the reaction kinetics transforms into Smith–Ewart phase 2 kinetics [50, 61]. Nomura et al. modeled continuous styrene emulsion polymerization and compared the model with experiments from two continuous stirred tanks in series [58]. In addition to models for the continuous emulsion polymerization of styrene, models for monomers with higher water solubility were developed [51, 62–65]. The product properties can be tuned by an optimized split dosing between the seed reactor and the main polymerization reactor [66]. Another interesting concept for overcoming oscillations was given by Barandiaran, who showed that miniemulsion polymerization avoids oscillations in a single CSTR [67, 68].

Furthermore, the density difference between continuous and dispersed phases should be kept in mind. Thus, a key factor for emulsion polymerization is proper dispersion of the different phases. For this task, pioneers in emulsion polymerization used the stirrer of the reaction vessel itself [69, 70]. Currently, at the industrial scale, the on-line preparation of preemulsions is preferred [71]. To dose stored preemulsions is challenging with respect to product quality, although the solid content of a stored preemulsion can be changed by partial creaming [72].

If there is a tendency for gelation at high conversions, a continuous stirred tank concept could have advantages over batch and **continuous tubular processes** [73–75]. The continuous emulsion polymerization of butadiene rubber is a typical example, with a train of up to15 CSTRs [37, 66, 73, 74, 76, 77]. Nowadays, such a large number of tanks in series is unusual because of the capital investment, but explainable by the long history of this product. The tank in series concept (in a modified sense) is highly attractive to chemical engineers. An example is the preparation of toner using high-g technology. Here, a three-stage spinning disc cascade is used for continuous emulsion polymerization [78]. In the context of process intensification, Asua reported the continuous production of a vinylacetate VeoVa10 in a train of two CSTRs [79].

# 3  Stirred Tubular Reactors

Stirred tubes have long been used to make emulsions. Walsh [80] presented a stirred tube apparatus that used pulses to improve the emulsifying efficiency. Later, this reactor concept was introduced into emulsion polymerization. An early example from 1940 for poly(vinyl chloride) (PVC) emulsion polymerization was presented by Jacobi [81]. Stirred tubular reactors are of special interest because they produce narrow residence time distributions, comparable to a stirred tank cascade in one pot, with one stirring device. Additionally, stirred tubular reactors allow back-mixing [82–84].

A current example is a stirred tubular reactor with different stirrer speeds along the tube [85]. Nörnberg investigated different stirrer geometries in a continuous tubular emulsion reactor [86], determining heat transfer characteristics and comparing vertical and horizontal setups.

# 4  Taylor–Couette Reactors

A Taylor reactor consists of an inner and outer cylinder, with a gap between. When the gap width is appropriate, the gap is filled with medium and the inner cylinder speed is high enough that so-called counter-rotating Taylor vortices are formed [87]. In principal, this reactor looks like a rheometer with cylindrical geometry. Thus, it is not surprising that Couette (1890) and Mallock (1896) were the first observers of vortices in that geometry. In 1923, Taylor published a mathematical model for this kind of reactor and successfully compared his calculations with own experiments [88]. Fage presented the first experiments and simulations on a continuous Taylor device. He determined the influence of axial flow on the appearance of Taylor vortices [89]. At present, this geometry is of special interest to researchers investigating and simulating flow phenomena. Grossmann gives a comprehensive review, presenting a huge number of different flow patterns [90]. Schmidt [91] presented flow simulations in combination with chemical reaction. The simulation showed fluctuations in concentration, which was demonstrated by Conrad experimentally for an organ pipe-like Taylor–Couette reactor [87].

The Taylor–Couette reactor was first used for liquid–liquid contactors [92]. Bernstein cited a very early work in this field [93]. From a chemical engineering viewpoint, the tunable flow characteristics make the continuous Taylor–Couette reactor interesting. By increasing the rotational speed of the inner cylinder, the back-mixing increases. Thus, the number of tanks in series is tunable [94, 95]. Kataoko attracted attention to the plug flow residence time distribution characteristics of this potential chemical reactor [96]. However, a low rotational speed resulted in axial plug flow characteristics with low inner vortex mixing [97]. Later, the heat transfer characteristics were investigated by Kataoka et al. [98], who showed the strong influence of vertical flow on heat transfer. In 1977, a Taylor–Couette reactor was patented by the Dow chemical company for the generation of a polymeric dispersion [99]. A first study of scale-up was published for dispersion processes in 1981 [100]. The authors designed a new dimensionless dispersion number. Today, the dependence of the Nusselt number on the Taylor number is known [101]. The specific heat transfer is in a similar range as for a CSTR.

For emulsion polymerization, this reactor concept became interesting through the works of Nomura [102, 103] and was patented for continuous emulsion polymerization [104]. Kataoka observed reduced coagulation [105] and Nomura proposed this reactor concept for shear sensitive lattices [106]. An experimental comparison between a Taylor reactor and different tubular reactors was given by Schmidt [91]. Asua showed the advantages and high flexibility of this reactor concept in comparison to CSTRs for high solid content emulsion polymerization [107]. In many cases, polymerization reactions cause changes in the rheological properties during the course of the reaction (i. e., increase in viscosity and change from Newtonian to shear thinning behavior). In the case of an increase in viscosity of the reaction medium, conical or organ pipe-like Taylor reactors were developed to ensure favorable tubular residence time characteristics [87, 108–111].

Continuous emulsion polymerization Talyor–Couette reactor concepts are vertical concepts. Langenbuch observed increasing reactor fouling with increased solid content and supposed that density difference was a cause of creaming [95]. In vertical operation mode, the stirring power input in the vertical direction is not high enough to avoid creaming [101]. Muessig was able to significantly reduce reactor fouling by operating the Taylor reactor in horizontal mode. For low solid content (6 wt%), Ruettgers showed high conversion for residence times of less than 1 min [112]. It is well accepted that a Taylor–Couette reactor is a slow mixer. Kind measured macromixing times in the range of 1 min [113]. However, Babar showed that, in the Taylor–Couette reactor with 15 min mean residence time, additional dosing is possible and advantageous with respect to high conversions [114]. To reduce shear with increased turbulence, an improved biotechnology Taylor–Couette-reactor has been developed, which is still to be investigated in continuous emulsion polymerization [115].

# 5  Tubular Reactors for Emulsion Polymerization

With the increasing industrial production of synthetic emulsion polymers, continuous tubular reactors are of practical interest. Early examples are documented from Du Pont and IG Farbenindustrie [9, 10]. An early example for volatile monomers is a tubular reactor from Goodrich [11]. Hydrostatic pressure was used to decrease the vapor pressure in a bent tubular reactor. The main advantage of tubular reactors is the high surface-to-volume ratio in comparison with stirred tank reactors, which is equivalent to a high cooling capacity. In this context, note that bending a tube improves the heat transfer performance [116]. Thus, tubular reactors allow higher reaction rates than tank reactors under safe conditions at comparable reactor volumes. Furthermore, the burst strength of tubular reactors is notable [117]. Adiabatic batch processes can be easily transferred into tubular processes [118, 119].

However, the large surface-to-volume ratio in comparison to stirred tank reactors is disadvantageous with respect to fouling [120] and fouling is a universal problem, not only in continuous emulsion polymerization [121, 122]. One way to avoid this is to use fouling-resistant reactor materials or to coat the reactor surface with a fouling-resistant material such as PTFE [123, 124]. Furthermore, active cleaning by a stirrer has been patented [125]. Another way to overcome the fouling problem is to use cleaning pigs, which are well established in pipeline transportation. For crude oil transportation through pipelines, complicated pigs have been described [126]. For continuous emulsion polymerization in tubular reactors, there are several patented examples that deal with cleaning pigs [127, 128]. Also, fully automatized cleaning pig setups are known [129, 130]. However, the cause of fouling is not completely understood [131]. McKenna stated that monomer droplets are a main factor in reactor fouling in continuous emulsion polymerization [132]. Further research on polymer fouling is necessary to make tubular reactors more popular for industrial emulsion polymerization. With improved in-line sensors (e. g., for turbidity and inductive conductivity), further progress in fouling research is expected [133–136].

Typically, continuous emulsion polymerization in tubular reactors is operated in a nonturbulent mode, otherwise tubular reactors become long and expensive. Use of bundled tubes is risky with respect to fouling. When the pressure loss differs between single tubes, the residence time can be become broader with time. A rare example on this topic is a work by Shork and Guo [42]. Furthermore, in laminar mode, creaming and sedimentation must be prohibited. Both are driven by the density difference between continuous and dispersed phases. An opportunity to overcome the creaming challenge is the implementation of static mixers [44, 137, 138]. However, static mixers bring additional surface into the reactor and increase the probability of fouling. Thereby, coiled and bent tubes are of special interest because of superimposed secondary flow.

Eustice observed the different flow characteristics in straight and coiled tubes [139]. Based on his experiments, Dean derived a mathematical model to calculate the upcoming counter-rotating vortices in coiled tubes [140]. As early as 1937, DuPont claimed a continuous tubular reactor with coils for high solid emulsion polymerization [10]. They described the advantages of this reactor with respect to coagulation and fouling. At present, it is well accepted that counter-rotating vortices not only increase laminar mixing efficiency, but also narrow residence time distribution and intensify heat transfer. Saxena showed that banding is advantageous for further narrowing the residence time distribution [141, 142]. Thus, coiled and bent tubes are widely used for high-performance liquid chromatography [143].

Moritz further improved the concept of coiled and bent reactors by developing a basket-type tubular reactor. The main characteristic of the basket-type tubular reactor is a very narrow residence time distribution compared with bent tubular reactors [91, 144]. Nevertheless, the construction enables the reactor to be cleaned by pigs. Moritz and Schmidt observed instationarities for the dependency of the Bodenstein number on the Reynolds number at very low Reynolds numbers. Horn could experimentally declare the instationarities by dead volumes [145].

Patent literature shows some industrial interest in tubular reactors. Continuous adiabatic processes have become more interesting [118]. Furthermore, coiled tubular reactors are of interest in millifluidic devices [72, 146, 147]. McKenna showed the feasibility of high solid continuous production of core–shell particles when a side feed is added to a coiled tube reactor [148].

# 6 Loop Reactors

The main characteristics of loop reactors are recycling of the reaction mass and high surface-to-volume ratio in comparison with CSTRs [149, 150]. If the recycle ratio is high enough, loop reactors have the same residence time distribution as CSTRs. Therefore, this kind of reactor is chosen for fast reactions that require continuous stirred tank residence time characteristics [151]. Furthermore, loop reactors are more flexible than CSTRs or tubular reactors because the residence time characteristics can be adjusted by tuning the recycle ratio. Thus, continuous loop concepts for making emulsions date back more than 90 years [152], but only recently started to be used in industrial emulsion production [30]. For high recycle ratios, property oscillations are observable, as in CSTRs [153, 154].

In academic research, batch and continuous loop reactors for emulsion polymerization are being intensively investigated. The highest polymerization rate was observed at the point of laminar–turbulent transition flow [155], which could be an indication of mass transport limitations in emulsion polymerization under special conditions, as reported by Smith [17]. Lee presented a loop reactor for emulsion polymerization of styrene with low solid content [156, 157]. In contrast, Asua and Pinto performed high solid content experiments and showed the suitability of loop reactors under industrial conditions [158]. The same authors also presented a comprehensive model for this complex process [159].

However, there are only a few documented industrial examples of the use of loop reactors in emulsion polymerization. Recently, Celanese seems to be paying special interest to this reactor type [160]. Examples are known from the coatings industry where vinyl acetate homopolymers and copolymers are produced in loop reactors [154, 159]. Further examples on continuous emulsion polymerization are given in Table 1.

**Table 1** Examples of continuous emulsion polymerization in loop reactors

| Patent author | Dispersed system | Company | Reactor |
|---|---|---|---|
| Adams [161] | Emulsion polymerization | Crown Brands Limited, Lancashire (GB) | Coiled loop followed by a coiled tube tubular reactor |
| Adams [162] | Emulsion polymerization | Reed Internationl PLC (GB) | Loop followed by tube with static mixers |
| Hopkins [163] | Inverse emulsion polymerization | The Lubrizol Cooperation (USA) | Coiled tube loop reactor |
| Eisenlauer et al. [164] | Emulsion or suspension polymerization | Wacker Chemie AG (D) | Jet loop reactor |
| Dowding et al. [165] | Suspension polymerization of porous particles | School of Chemistry, University of Bristol (GB) | Coiled PTFE tube reactor |
| Herrle and Beckmann [166] | Emulsion polymerization | BASF AG (D) | Straight tubes with straight post polymerization tube |
| Gonzalez et al. [167] | Copolymerization | Polymat (ES) | Straight tubes |
| de la Cal et al. [149] | Terpolymerization | Polymat (ES) | Straight tubes |

In loop reactors, shearing and coagulation of the boosted emulsion inside the pumping device is a challenge. To overcome problems with the boosting device, stirrers were introduced as boosters. Tanaka presented a circular loop reactor, which was intensively investigated for suspension polymerization [168]. The main disadvantage of this concept is the increasing centrifugal force from the inner to the outer part of the tube cross-section, which results in a bend curvature and material separation across the circular tube [169, 170].

# 7   Pulsed Reactors

The advantage of pulses for dispersing tasks has been known since the beginning of polymer reaction engineering [80]. Some very complicated machines have been constructed, such as having up to six pulse generators in one tube [171]. Mass transport could be of importance in speeding up continuous emulsion polymerization. Although oscillations increase heat and mass transfer, this type of reactor is of industrial interest and not only for continuous emulsion polymerization [172, 173]. Additionally, pulsed continuous tubular reactors can shorten process development times [174].

Extensive work in the field of continuous emulsion polymerization began at the end of the 1980s. DSM patented a pulsed packed column for continuous emulsion polymerization [175–177]. Ray observed the advantages of pulses with respect to narrowing the residence time distribution and for reducing wall fouling [39]. Meuldijk investigated continuous pulsed sieve columns and compared them with batch and CSTRs, showing the advantages of a pulsed packed column with respect to heat transfer [33]. With added side streams, products are comparable to those obtained in a semibatch reactor [178]. More recently, Giudici used a model with a precalculated side stream dosing profile in order to compensate for copolymerization composition drift [179].

# 8 Continuous Heterophase Polymerization in Micro- and Mesoscale Reactors

Microreaction technology ideas in heterophase polymerization are older than the buzzword "microreaction technology" [180–182]. But, first it was necessary to develop the technologies to produce microstructured devices. Thus, there are now many ideas on how to introduce microreaction technology into polymer reaction technology [183]. A recent introduction to microreaction particle generation was given by Kumacheva and Garstecki [184]. Several successful attempts to use microreaction technology for particle generation were made at the laboratory scale. Of special interest are new and advanced particle generation technologies such as membrane emulsification [185], coflow technologies [147, 186], melt emulsification [187], and micromixers for emulsification [188, 189]. Mongeon compared an oscillatory flow reactor with microreaction devices and showed that microreactors significantly improve the mass transfer in dispersed systems [172]. Nevertheless, widespread commercial use for commodity production seems to be far in the future. An industrial example in use is the production of carrier particles for ion exchange in an upscaled microreaction process [190]. A few other patents describe similar processes [191, 192].

Mixing is another big field of microreaction technology. Micromixers are especially advantageous for continuous emulsion polymerization, where the monomer must be distributed on-line at the micrometer scale with high reproducibility [193]. The mixers are connected to the polymerization reactor. An interesting example is the combination of a traditional micromixer with a mesoscale bent tubular reactor for continuous emulsion copolymerization [72].

The surface-to-volume ratio of microreactors is much higher than in conventional tubular reactors. Pressure loss is also higher because of the small reactor diameter, which is the reason for the small number of publications dealing with microreactors for continuous emulsion polymerization. However, microreactors are established for high value products. An example is Grignard metathesis polymerization of 3,4-ethylenedioxythiophene [194], where the advantage of the high cooling capacity is used.

# 9 Conclusion

Recently, the research field of continuous emulsion polymerization has become highly dynamic and attractive. Extended research to elucidate the mechanism of fouling is necessary and challenging. Furthermore, model controlled polymerizations offer many opportunities for scientific activity. Different continuous reactor concepts are available and have been scientifically characterized, offering many opportunities to develop and manufacture innovative products. However, this requires implementing continuous reactor concepts early in the research stages, not only at the industrial stage. For this, additive manufacturing can work as a door opener, because it offers the opportunity to fabricate cheap research reactors. Last but not least, reactor concepts developed for other products can be adapted to the field of continuous emulsion polymerization.

# References

1. Tauer K, Antonietti M (2003) 90 years of polymer latexes and heterophase polymerization: more vital then ever. Macromol Chem Phys 204(2):207–219
2. Ditmar R (1912) Die Synthese des Kautschuks. Mitteilungen des naturwissenschaftlichen Vereins für Steiermark 48:435–449
3. Staudinger H, Frey H, Starck K (1927) Verbindungen hohen Molekulargewichts IX. Polyvinylacetat und Polyvinylalkohol. Ber Dtsch Chem Ges B 60B:1782–1792
4. Tannehill AL (1919) Process for producing arificial resin. Patent US 1389791
5. Bayer FVF (1913) Verfahren Darstelllung von künstlicheKautschuk. Patent CH 64857
6. Siemens-Schuckertwerke (1928) Verfahren zur Polymerisation von Kondensationsprodukten aus mehrwertigen Alkohlen und mehrbasichen Säuren. Patent DE 572125
7. Miller SP (1923) Mixing machine and process. US1670593
8. Bannon LA, Roselle NJ (1936) Method of producing polymers of high molecular weight. Patent US 2317878
9. Anon (1937) Improvements in or relating to the manufacture of synthetic rubber-like masses. Patent GB517951
10. Marks BM (1937) Process of polymerization. Patent US2161481
11. Schoenfeld FK, Semon WL (1938) Method for continuous polymerization. Patent US2259180A
12. Calcott WS, Woodstown NJ, Starkwether HW (1939) Continuous process for preparing dispersions of polymerized elastogenic dienes. Patent US2394291A
13. Seebold JE (1946) Ethylene polymeriaztion process. Patent US2575643
14. Dunlop RD, Reese FE (1948) Continuous polymerization in Germany. Ind Eng Chem 40(4):654–660
15. Harkins WD (1945) A general theory of the reaction loci in emulsion polymerization. J Chem Phys 13(9):381–382
16. Harkins WD (1950) General theory of mechanism of emulsion polymerization. J Polym Sci 5(2):217–251
17. Smith WV (1948) The kinetics of styrene emulsion polymerization. J Am Chem Soc 70(11):3695–3702
18. Smith W, Ewart RH (1948) Kinetics of emulsion polymerization. J Chem Phys 16:592–599
19. Elvers B (2011) Ullmann's encyclopedia of industrial chemistry. Wiley-VCH, Weinheim

20. Mark HF, Seidel A (2014) Encyclopedia of polymer science and technology, vol 15. 4th edn. Wiley, Hoboken
21. Asua JM (2007) Polymer reaction engineering. Blackwell, Oxford
22. Chern C-S (2008) Principles and applications of emulsion polymerization. Wiley, Hoboken
23. Erbil YH (2000) Vinyl acetate emulsion polymerization and copolymerization with acrylic monomers. CRC, Boca Raton
24. Gilbert RG (1995) Emulsion polymerization: a mechanistic approach. Academic, London
25. Lovell PA, El-Aasser MS (1997) Emulsion polymerization and emulsion polymers. Wiley, Chichester
26. Soares JBP, McKenna TFL (2013) Polyolefin reaction engineering. Wiley-VCH, Weinheim
27. van Herk A (2013) Chemistry and technology of emulsion polymerisation2nd edn. Wiley, Chichester
28. Sheibat-Othman N, Vale HM, Pohn JM, McKenna TFL (2017) Is modeling the PSD in emulsion polymerization a finished problem? An overview. Macromol React Eng 27:1600059
29. Asua JM (2004) Emulsion polymerization: from fundamental mechanisms to process developments. J Polym Sci A Polym Chem 42:1025–1041
30. Asua JM (2016) Challenges and opportunities in continuous production of emulsion polymers: a review. Macromol React Eng 10:311–323
31. Choi KY (2016) Continuous processes for radical vinyl polymerization. In: Mishra MK, Yagci Y (eds) Handbook of vinyl polymers. CRC, Boca Raton, pp 347–368
32. Leiza JR, Meuldijk J (2013) Emulsion copolymerisation, process strategies. In: Herk AV (ed) Chemistry and technology of emulsion polymerisation2nd edn. Wiley, Chichester, pp 75–104
33. Meuldijk J, Mayer MJJ, Thoenes D (1994) Emulsion polymerization in various reactor types: recipies with high monomer content. Chem Eng Sci 49(24B):4971–4980
34. Nomura M, Tobita H, Suzuki K (2005) Emulsion polymerisation: kinetic and mechanistic aspects. Adv Polym Sci 175:1–128
35. Penlides A, Dube MA, Soares JBP, Hamielec AE (1997) Mathematical Modeling of multicomponent chain-Groth polymerization in batch, semibatch and continuous reactors: a review. Ind Eng Chem Res 36:966–1015
36. Phoelein GW (1997) Reaction engineering for emulsion polymeriaztion. In: Asua JM (ed) Polymeric dispersions: principles and applications. Kluwer Academic, Dordrecht, pp 305–332
37. Phoelein GW, Dougherty DJ (1977) Continuous emulsion polymerization. Rubber Chem Technol 50(3):601–638
38. Poehlein GW (1983) Emulsion polymerization in continuous reactors. Science and technology of polymer colloids. NATO ASI series, vol 67–68. Springer Netherlands, Dordrecht, pp 112–139
39. Ray WH, Paquet Jr DA (1994) Tubular reactors for emulsion polymerization: I. Experimental investigation. AICHE J 1(40):73–87
40. Vale H, McKenn T (2005) Modeling particle size distribution in emulsion polymerization reactors. Prog Polym Sci 30(10):1019–1048
41. Li X et al. (2016) Progress in reactor engineering of controlled radical polymerization: a comprehensive review. React Chem Eng 1:23–59
42. Schork FJ, Guo J (2008) Continuous miniemulsion polymerization. Macromol React Eng 2:287–303
43. Alarcia F, de la Cal JC, Asua JM (2006) Process intensification in the production of specialty waterborne polymers. Macromol Mater Eng 291:428–437
44. Rossow K et al. (2016) Transfer of emulsion polymerization of styrene and n-butyl acrylate from semi-batch to a continuous tubular reactor. Macromol React Eng 10:324–338

45. Bashir MA, McKenna TF (2017) Reaction engineering of polyolefins: the role of catalyst supports in ethylene polymerization on metallocene catalysts. Adv Polym Sci doi: 10.1007/12_2017_23
46. Rupaner R et al (1998) Use of a single-stage or multistage stirrer to prepare polymers. Patent US6252018B1
47. Winding CC (1948) Polymerization. Ind Eng Chem 40(9):1643–1649
48. Logemann H (1961) Allgemeines zur Technik der Polymerisation in heterogener Phase. Houben-Weyl methods of organic chemistry, vol XIV/1. 4th edn. Thieme, Stuttgart, p 154
49. Beckmann G, Matis H (1966) Simulation einfacher Vorgange in Reaktoren. Chem Ing Tech 38(3):209–214
50. DeGraff AW, Poehlein GW (1971) Emulsion polymerization of styrene in a single continuous stirred-tank reactor. J Polym Sci A-2 9:1955–1976
51. Gerrens H, Kuchner K (1970) Continuous emulsion polymerisation of styrene and methyl methacrylate. Br Polym J 2:18–24
52. Fikentscher H (1937) Verfahren zur kontinuierlichen Emulsionspolymerisation. Patent DE 900019
53. Ray WH, Villa CM (2000) Nonlinear dynamics found in polymerization processes – a review. Chem Eng Sci 55(2):275–290
54. Brooks BW (1997) Why are polymerization reactors special? Ind Eng Chem Res 36:1158–1162
55. Omhura N, Kataoka K, Watanabe S, Masayoshi O (1998) Controlling particle size by self-sustained oscillations in continuous emulsion polymerization of vinyl acetate. Chem Eng Sci 53(12):2129–2135
56. Wall FT, Delbecq CJ, Florin RE (1952) Studies in continous polymerization. J Polym Sci 9 (2):177–192
57. Wilson WK (1948) Continuous emulsion polymerization. Patent US2587562
58. Nomura M et al. (1971) Continuous flow operation in emulsion polymerization of styrene. J Appl Polym Sci 15(3):675–691
59. Omi S, Ueda T, Kubota H (1969) Continuous operation of emulsion polymerization of styrene. J Chem Eng Jpn 2(2):193–198
60. Ueda T, Omi S, Kubota H (1971) Experimental study of continuous emulsion polymerization of styrene. J Chem Eng Jpn 4(1):50–54
61. Antonucci JP, Taylor MA, Takamura K, Racz R (2001) Active small diameterseed latex for continuous emulsion polymerization. Patent US 2004/0106725
62. Kiparissides C, MacGregor JF, Hamielec AE (1979) Continiuous emulsion polymerization. Modeling oscillations in vinyl acetate polymerization. J Appl Polym Sci 23:401–418
63. Kiparissides C, MacGregor JF, Hamielec AE (1980) Continuous emulsion polymerization of vinyl acetate. Part I: Experimental studies. Can J Chem Eng 58(1):48–55
64. Penlidis A, MacGregor JF, Hamielec AE (1985) Dynamic modeling of emulsion polymerization reactors. AIChE J 31(6):881–889
65. Schork FJ, Ray WH (1987) The dynamics of the continuous emulsion polymerization of methylmethacrylate. J Appl Polym Sci 34:1259–1276
66. Penlidis A, MacGregor JF, Hamielec AE (1989) Continuous emulsion polymerization: design and control of CSTR trains. Chem Eng Sci 44(2):273–281
67. Barandiaran MJ, Aizpurua I (1999) Comparison between conventional emulsion and miniemulsion polymerization of vinyl acetate in a continuous stirred tank reactor. Polymer 40:4105–4115
68. Barandiaran M, de la Cal J, Amalvy JI, Aizpurua I (2001) High solids content miniemulsion polymerization of vinyl acetate in a continuous stirred tank reactor. Polymer 42:1417–1427
69. Mast WC, Fisher CH (1949) Emulsion polymerization of acrylic esters and other vinyl monomers. Ind Eng Chem 41(4):790–797
70. Renfrew A, Gates WEF (1938) Polymerization of water insoluble organic compounds dispersed in an aqueus vehicle. Patent US 2296403

71. Keller A et al (1997) Method for the production of a polymer dispersion by radical aqueous emuslion polymerization with a continuously produced aqueous monomer emulsion. Patent CA 2272863 C
72. Retusch C (2013) Reaktionstechnische Untersuchungen zur schnellen kontinuierlichen Emulsionspolymerisation im multiskaligen Milli-Reaktor. Wissenschaft & Technik, Berlin
73. Minari RJ, Gugliotta LM, Vega JR, Meira GR (2006a) Continuous emulsion styrene-butadiene rubber (SBR) process: computer simulation study for increasing production and for reducing transients between steady states. Ind Eng Chem Res 45(1):245–257
74. Minari RJ, Gugliotta LM, Vega JR, Meira GR (2006b) Emulsion copolymerization of acrylonitrile and butadiene in a train of CSTRs. Intermediate addition policies for improving the product quality. Lat Am Appl Res 36(4):301–308
75. Steffers F, Bochmann D, Zill W (1989) Verfahren zur kontinuierlichen Redoxcopolymerisationvon Butadien in wäßriger Lösung. Patent DE D272468
76. Madhuranthakam CMR, Penlidis A (2016). Processes 4(1):6
77. Musch R et al (1986) Process for polymerizing chloroprene. Patent EP0235636B1
78. Lai Z, Cheng C-M, Wolfe CM, Jackson MA (2005) Latex emulsion polymerization in spinnig disc reactors or rotating tubular reactors. Patent US7683142
79. Agirrea A, Weitzel H-P, Hergeth W-D, Asua JM (2015) Process intensification of VAc–VeoVa10 latex production. Chem Eng J 266:34–47
80. Walsh VG (1929) Emulsifying an analogous apparatous. Patent US 1780853
81. Jacobi B (1952) Zur Kolloidchemie der Emulsionspolymerisation. Angew Chem 64:539–543
82. Ghosh S, Moritz H-U, Reichert K-H (1983) Verweilzeitverhalten gerührter Strömungsrohre. Chem Ing Tech 55(8):635–638
83. Moritz H-U (1982) Kontinuierliche Perlpolymerisation von Vinylacetat in einem gerührten Rohrreaktor. PhD thesis, Berlin
84. Moritz H-U, Reichert K-H (1981) Zur kontinuierlichen Perlpolymerisation im gerührten Rohrreaktor. Chem Ing Tech 53(5):386–387
85. Marx RE, Nelson JM, Hanley KJ (2004) Stirred tube reactor and method of using the same. Patent US6969491 B1
86. Nörnberg C (2016) Kontinuierliche Emulsionscopolymerisation von Vinylacetat in einem gerührten Rohrreaktor. Wissenschaft und Technik, Berlin
87. Conrad I (2003) Beiträge zur Charakterisierung eines konischen Taylors zur kontinuierlichen Massepolymerisation. Wissenschaft und Technik, Berlin
88. Taylor GI (1923) Stability of a viscous liquid contained between two rotating cylinders. Philos Trans R Soc Lond A 223:289–343
89. Fage A (1938) The influence of wall oscillations, wall rotation, and entry eddies, on the breakdown of laminar flow in an annular pipe. Proc R Soc Lond A Math Phys Sci 165 (923):501–529
90. Grossmann S, Lohse D, Sun C (2015) High–Reynolds number Taylor-Couette turbulence. Annu Rev Fluid Mech 48:53–80
91. Schmidt W (1998) Entwicklung von REaktoren für die kontinuierliche Emulsionspolymerisation als Alternative zum Durchfluß-Rührkessel und zur Durchflu-ß-Rührkesselkaskade. Wissenschaft & Technik, Berlin
92. Bernstein GJ, Grosvenor DE, Levitz NM, Lenc JF (1973) A high-capacity annular centrifugal contactor. Nucl Technol 20(3):200–202
93. Clay PH (1940) The mechanism of emulsion formation in turbulent flow. Akademie van Wetenschappen, Amsterdam, pp 852–965
94. Crookewit P, Honig CC, Kramers H (1955) Longitudinal diffusion in liquid flow through an annulus between a stationary outer cylinder and a rotating inner cylinder. Chem Eng Sci 4:111–118
95. Langenbuch J (2001) Thermisches Beherrschen schneller kontinuierlicher Emulsionspolymerisationen. Wissenschaft & Technik, Berlin

96. Kataoka K, Doi H, Hongo T, Futagawa M (1975) Ideal plug-flow properties of Taylor vortex flow. J Chem Eng Jpn 8(6):472–476
97. King GP, Takeda Y (1999) Characterizing the Taylor-Couette reactor. In: Book of abstracts 2nd international symposium on ultrasonic doppler methods for fluid mechanics and fluid engineering, 20–22 Sept 1999, pp 11–12
98. Kataoka K, Doi H, Komai T (1977) Heat/mass transfer in Taylor vortex flow with constant axial flow rates. Int J Heat Mass Transf 20(1):57–63
99. Warner GL, Leng DL (1977) Continuous process for preparing aqueous polymer microsuspensions. Patent US4123403 A
100. Leonard RA, Bernstein GJ, Pelto RH, Ziegler AA (1981) Liquid-liquid dispersion in turbulent couette flow. AICHE J 27(3):495–503
101. Müssig S (2005) Prozessintensivierung der Emulsionspolymerisation: Der kontinuierliche Taylor-REaktor. Wissenschaft & Technik, Berlin
102. Imamura T, Saito K, Ishikura S, Nomura N (1991) A new approach to continuous emulsion polymerization. In: Preprints of international symposium on polymeric microspheres, pp 151–154
103. Imamura T, Saito K, Ishikura S, Nomura M (1993) A new approach to continuous emulsion polymerization. Polym Int 30:203–206
104. Imamura T et al (1995) Continuous polymerization method and apparatus. Patent JP EP0498583A
105. Kataoka K et al. (1995) Emulsion polymerization of styrene in a continuous Taylor vortex flow reactor. Chem Eng Sci 50(9):1409–1416
106. Wei X, Takahashi H, Sato S, Nomura M (2001) Continuous emulsion polymerization of styrene in a single Couette-Taylor vortex flow reactor. J Appl Polym Sci 80:1931–1942
107. Gonzales G et al. (2009) Production of widely different dispersed polymers in a continuous Taylor–Couette reactor. Macromol React Eng 3:233–240
108. Jung W-A, Rink H-P, Meinecke H, Krull J (2001) Method for continuous polymerizing in mass quantities and Taylor reactor for carrying out this method. Patent WO03/031056
109. Kossak S (2000) Kontinuierliche Lösungspolymerisation von Acrylmonomeren in Taylor-Reaktoren1st edn. Wissenschaft und Technik, Berlin
110. Moritz H-U et al (1998) Taylorreaktor für Stoffumwandlungen, bei derenVerlauf eine Änderung der Viskosität des Reaktionsmediums eintritt. Patent DE19828742
111. Moritz H-U et al (1999) Verfahren zur Polymerisation olefinisch ungesättigter Monomere mittels eines Taylorreaktors. Patent DE19960389
112. Rüttgers D, Negoita I, Pauer W, Moritz H-U (2007) Process intensification of emulsion polymerization in the continuous Taylor reactor. Macromol Symp 259:26–31
113. Racina A, Liu Z, Kind M (2010) Mixing in Taylor-Couette flow. In: Mewes D et al. (eds) Micro and macro mixing – analysis, simulation and numerical calculation. Springer, Berlin Heidelberg, pp 125–139
114. Babar M (2013) Continuous emulsion copolymerization of vinyl acetate and VeoVa 10 using Taylor reactor. Wissenschaft und Technik, Berlin
115. Sorg R et al. (2011) Minimizing hydrodynamic stress in mammalian cell culture through the lobed Taylor-Couette bioreactor. Biotechnol J 6:1504–1515
116. Baurmeister U, Brauer H (1979) Laminare Strömung und Wärmeübergang in Rohrwendeln und Rohrspiralen. VDI-Forschungsheft 593 ed. VDI, Düsseldorf
117. Roedel MJ (1951) Ethylene/vinyl acetate polymerization process. Patent US2703794A
118. Kane J, Durham JF III (2015) Continuous adiabatic inverse emulsion polymerization process. Patent US9434793B1
119. Shoaf GL, Poehlein GW (1989) Batch and continuous emulsion copolymerization of ethyl acrylate and methylacrylic acid. Polym-Plast Technol Eng 28(3):289–317
120. Willeke F (2000) Vorrichtung zur Polymerisation in einem Schalufenreaktor. s.l. Patent DE 10037153 C2
121. Irvin HB (1962) Bulk polymerization. Patent US3182050

122. Platzer N (1970) Design of continuous and batch. Ind Eng Chem 62(1):6–20
123. Fairchok WJ, Stanley FW (1980) Continuous polymerization of water-miscible monomers. Patent US4331787
124. Sütterlin N, Blitz H-D, Jagsch K-H, Tessmer D (1982) Kontinuierliches Emulsionspolymerisationsverfahren. Patent DE3222002A1
125. Feller R et al (2010) Reactor and method for continiuous polymerization. Patent US 2012/0302706 A1
126. Sharma MM (2015) Electromagnetic pig for oil and gas pipelines. Patent WO2016/186919
127. Adams DC (2006) Loop reactor for polymerization and method for cleaning thereof. Patent EP1973647B1
128. Adams DC, Jones H, Geddes KR (1998) Closed loop continuous polymerization reactorand polymerization process. Patent EP1113871B1
129. Adams D (2012) Continuous emulsion polymerization reactor and pigging system. Patent WO2014/085248
130. Gairns SA, Joustra J (1997) Apparatus for the self-cleaning of process tubes. Patent US5890531
131. Urrutia J, Peña A, Asua JM (2017) Reactor fouling by preformed latexes. Macromol React Eng 11:1–14
132. Ouzineb K, Graillat C, McKenna T (2004) Continuous tubular reactors for latex production: conventional emulsion and miniemulsion polymerizations. J Appl Polym Sci 91:2195–2207
133. Bloch D, Bröge P, Pauer W (2017) Inline turbidity measurements of batch emulsion polymerization. Macromol React Eng doi:10.1002/mren.201600063
134. Santos AF et al. (2003) Online monitoring of the evolution of the number of particles in emulsion polymerization by conductivity measurements. I. Model formulation. J Appl Polym Sci 90:1213–1226
135. Tabrizi FF, Mahdi Fadaee M (2017) Experimental investigation of continuous conductivity measurements during emulsion polymerizations having unstable/high-fouling reaction mixture. J Appl Polym Sci 134:44446
136. Zhao F et al. (2013) Online conductivity and stability in the emulsion polymerization of n-butyl methacrylate: batch versus semibatch systems. J Appl Polym Sci 130:4001–4013
137. Lüth FG (2015) Experimentelle Untersuchungen und numerische Simulationen eines Smart Scale Prozesses zur Intensivierung der Emulsionspolymerisation1st edn. Wissenschaft und Technik Verlag, Berlin
138. Schmidt J (2011) Kinetische und reaktionstechnische Untersuchungen zur kontinuierlichen schnellen Emulsionscopolymerisation1st edn. Wissenschaft und Technik Verlag, Berlin
139. Eustice J (1911) Experiments on stream-line motion in curved pipes. Proc R Soc Lond Ser A 85(576):119–131
140. Dean WR (1927) XVI. Note on the motion of fluid in a curved pipe. Philos Mag J Sci 20 (4):208–223
141. Kockmann N, Kurt SK, Gelhausen MG (2015) Axial dispersion and heat transfer in a milli/microstructured coiled flow inverter for narrow residence time distribution at laminar flow. Chem Eng Technol 38(7):1122–1130
142. Saxena AK, Nigam KDP (1984) Coiled configuration for flow inversion and its effect on residence time distribution. AICHE J 30(3):363–368
143. Gjerde DT, Hanna CT (2003) Biomolecule open channel solid phase extraction systems and methods. Patent US2004/0224425
144. Moritz H-U et al (1996) Vorrichtung zur kontinuierlichen Durchführung chemischer Reaktionen. Patent DE19634450
145. Horn J (2005) Kontinuierliche Copolymerisation von Acrylaten in Masse1st edn. Wissenschaft und Technik, Berlin
146. Garagalza O et al. (2017) Droplet-based millifluidic device under microwave irradiation: temperature measurement and polymer particle synthesis. Chem Eng J 308:1105–1111

147. Kunowa K (2013) Inverse Suspensionspolymerisation wasserlöslicher Polymere mittels segmentierter Mikrofluidik. Wissenschaft und Technik, Berlin
148. Hordy N, McKenna TFL (2012) A continuous tubular reactor for core–shell. Can J Chem Eng 90:437–441
149. de la Cal JC, Abad C, Asua JM (1995) Emulsion copolymerization of vinyl esters in continuous reactors: comparison between loop and continuous stirred tank reactors. J Appl Polym Sci 56:419–424
150. Legrand J, Belleville P (2002) Flow characteristics and transport phenomena in toroidal loop reactors. Chem Eng Technol 25(6):667–670
151. Blenke H (1979) Loop reactors. Adv Biochem Eng 13:121–214
152. Kirschbaum L (1924) Apparatous for making bituminous emulsions. Patent US1560824
153. Bataille P, Poormahdian S (2000) Emulsion copolymerization in a tubular reactor. J Appl Polym Sci 75:833–842
154. Geddes KR (1989) Start-up and growth mechanisms in the loop reactor. Br Polym J 21:433–441
155. Rollin AL, Patterson I, Huneault R, Bataille P (1977) Continuous emulsion polymerisation of styrenein a tubular reactor. Can J Chem Eng 10(55):565–571
156. Lee D-Y, Kuo J-F, Wang J-H, Chen C-Y (1990) Study on the continuous loop tubular reactor for emulsion polymerization of styrene. Polym Eng Sci 30(3):187–192
157. Lee D-Y, Wang J-H, Kuo J-F (1992) The performance of the emulsion polymerization of styrene in a continuous loop tubular reactor. Polym Eng Sci 32(3):198–205
158. Araújo PHH et al. (1999) Emulsion polymerization in a loop reactor: effect of the operation conditions. Polym React Eng 7(3):303–326
159. Araújo PHH, de la Cal JC, Asua JM, Pinto JC (2001) Modeling particle sicze distribution (PSD) in emulsion copolymerization reactions in a continuous loop reactor. Macromol Theory Simul 10(8):769–779
160. Adams D (2014) Continuous emulsion polymerization process and initiation system. Patent US2016/0032023
161. Adams DC (2008) Loop reactor for emulsion polymerization. Patent US7858715B2
162. Adams DC (1983) Production for emulsion polymers. Patent EP0145325A2
163. Hopkins TR (1981) Method for production of inverse emulsion polymers. Patent US4419466
164. Eisenlauer J et al (2011) Verfahren zur Herstellung von Polymerisaten mittels Emulsions- oder Suspensionspolymerisation. Patent DE 102011005388 A1
165. Dowding PJ, Godwin JW, Vincent B (2000) Production of porous suspension polymers using a continuous tubular reactor. Colloid Polym Sci 278(4):346–351
166. Herrle K, Beckmann W (1965) Verfahren zur kontinuierlichen Emulsionspolymerisation olefinisch ungesättigter Verbindungen. Patent DE DD1240280
167. Gonzalez I, Paulis M, de la Cal JC, Asua JM (2008) Kinetic and microstructural study of the continuous emulsion polymerization of an all-acrylics formulation in the loop reactor. Chem Eng J 142(2):199–208
168. Tanaka M, Senda T, Hosogai K (1989) Flowing characteristics in a circular loop reactor. Chem Eng Res Des 67(4):423–427
169. Khalid A, Legrant J, Rosant JM (1996) Turbulent flow induced by an impeller in a closed toroidal loop. J Fluids Eng 118(4):677–684
170. Pauer W (1996) Reaktionstechnische Aspekte der Suspensionspolymerisation von Styrol1st edn. VDI, Düsseldorf
171. Brooks CH (1957) Olefin polymerization in a pulsating reactor. Patent US2927006A
172. Mongeon SS et al. (2016) Liquid–liquid mass transfer in an oscillatory-flow mesoscale coil reactor without baffles. Org Process Res Dev 20:733–741
173. Ni X et al. (2003) Mixing through oscillations and pulsations – a guide to achieve process enhancements in the chemical and process industries. Trans IChemE 81(Part A):373–383
174. McDonough JR, Phan AN, Harvey AP (2015) Rapid process development using oscillatory baffled mesoreactors – a state-of-the-art review. Chem Eng J 265:110–121

175. Hoedemakers GFM (1990) Continuous emulsion polymerization in a pulsed packed column. Dissertation, Technische Universiteit Eindhoven, Eindhoven
176. Hoedemakers GFM, Thoenes D (1990) Continuous emulsion polymerization in a pulsed packed column. In: Kleintjens LA, Lemstra PJ (eds) Integration of fundamental polymer science and technology, vol 4. Elsevier, New York, pp 182–193. https://link.springer.com/chapter/10.1007/978-94-009-0767-6_23
177. Hoedemakers GFM, Thoenes D, Van DLJLM (1988) Process for polymerization in a packed pulsating column. Patent NL EP0336469B1
178. Scholtens CA, Meuldijk J, Drinkenburg AAH (2001) Production of copolymers with a predefined intermolecular chemical composition distribution by emulsion polymerisation in a continuously operated reactor. Chem Eng Sci 56(3):955–962
179. Carvalho ACSM, Sayer C, Chicoma D, Giudici S (2010) In: ISCRE 21 proceedings [online]. www.iscre.org/iscre21/abstracts/207.pdf. Accessed 16 Nov 2016
180. Breedis J (1940) Method of polymerization. Patent US2326326
181. Hutchinson HM, Staudinger JJP (1946) Production of polystyrene beads. Patent US2566567A
182. Stearns RF, Elisabeth NJ (1949) Method and apparatus for continuous flow mixing. Patent US2645463
183. Hessel V, Serra C, Löwe H, Hadziioannou G (2005) Polymerisationen in mikrostrukturierten Reaktoren: Ein Überblick. Chem Ing Tech 77(11):1693–1714
184. Kumacheva E, Garstecki P (2011) Microfluidic reactors for polymer particles. Wiley, Chichester
185. Simons JMM, Keurentjes JTF, Meuldijk J (2013) Micron-sized polymer particles by membrane emulsification. Macromol Symp 333(1):102–112
186. Liu Z, Lu Y, Yang B, Luo G (2011) Controllable preparation of poly(butyl acrylate) by suspension. Ind Eng Chem Res 50:11853–11862
187. Köhler K, Hensel A, Kraut M, Schuchmann HP (2011) Melt emulsification – is there a chance to produce particles without additives? Particuology 9(5):505–511
188. Li Y et al. (2013) Development of a continuous emulsification process for a highly viscous dispersed phase using microstructured devices. Green Process Synth 2:499–507
189. Pennemann H et al. (2004) Flüssig/Flüssig-Dispergierung im Interdigital-Mikromischer. Chem Ing Tech 76(5):651–659
190. Lange P-M, Strüver W ( 1981) Verfahren zur Herstellung von Perlpolymerisaten einheitlicherTeilchengröße. Patent DE D0046535
191. Deppe H et al (2000) Polystyrene microspheres and a method for their production. Patent US 2004/0012105 A1
192. Petela G, Bleijenberg KC (2002) Controlled suspension polymerization process without mechanical agitation. Patent EP1408053
193. Lüttgen K (2011) Process for continuous emulsion polymerization. Patent US 20130123427 A1
194. Kumar A et al. (2014) Continuous-flow synthesis of regioregular poly(3-hexylthiophene): ultrafast polymerization with high throughput and low polydispersity index. J Flow Chem 4 (4):1

Adv Polym Sci (2018) 280: 19–64
DOI: 10.1007/12_2017_23
© Springer International Publishing AG 2017
Published online: 20 September 2017

# Reaction Engineering of Polyolefins: The Role of Catalyst Supports in Ethylene Polymerization on Metallocene Catalysts

**M. Ahsan Bashir and Timothy F.L. McKenna**

**Abstract** This chapter presents a brief look at different methods for the polymerization of ethylene using supported metallocene catalysts, then focuses on the effects that the properties of silica gel supports can have on catalyst behavior and the polymerization process. A review of the literature reveals that surprisingly little work has been done on the role of the support in polymerization, perhaps because of the numerous confounding issues that arise. Even less appears to have been done in terms of understanding how the support structure impacts catalyst formulation. More attention needs to be paid to controlling factors such as particle size and pore structure in studies meant to elucidate the role of the support in the polymerization process.

**Keywords** Catalyst formulation • Ethylene polymerization • Metallocene • Pore size • Porosity • Silica supports

## Contents

M. Ahsan Bashir
Chemistry, Catalysis, Polymers and Processes – UMR5265, CNRS, UCB-Lyon 1, ESCPE
Lyon, Université de Lyon, 43 Blvd du 11 Novembre 1918, 69616 Villeurbanne Cedex, France

Dutch Polymer Institute (DPI), P.O. Box 902, 5600 AX Eindhoven, The Netherlands

T.F.L. McKenna (✉)
Chemistry, Catalysis, Polymers and Processes – UMR5265, CNRS, UCB-Lyon 1, ESCPE
Lyon, Université de Lyon, 43 Blvd du 11 Novembre 1918, 69616 Villeurbanne Cedex, France
e-mail: timothy.mckenna@univ-lyon1.fr

# 1   Introduction

"Polyolefin" is a generic name given to the homopolymers of lighter olefins such as ethylene and propylene, and their copolymers with higher α-olefins (1-butene, 1-hexene, 1-octene, etc.). Polyethylenes (PEs) and polypropylenes (PPs) are the most widely produced families of polyolefins, and these can be further divided into sub-categories as a function of their composition and physical properties. Although the exact production figures for these two polymers vary depending on the source, it is clear that the combined production of PE and PP is about 150 million metric tons per year, making them the most widely manufactured polymers on the planet. This is despite the fact that they contain no special functional groups and are made only of carbon and hydrogen atoms. Belying their apparently simple structure, the molecular architecture of polyolefins can be tailored in such a way that one can exercise a great deal of control over their final physical properties; PE can be used for applications ranging from food wraps to body armor. PPs range from toughened materials for use as garden chairs to high-performance elastomers. This fine control over the properties of apparently simple molecules lies in the smart combination of chemistry and process, in other words, through judicious application of polymer reaction engineering.

# 2   A Brief Look at Polyolefins

The most important ethylene-based polyolefins are often classified according to density, which depends on the degree of branching of the macromolecules. As illustrated in Fig. 1, PEs include low density polyethylene (LDPE), linear low density polyethylene (LLDPE), and high density polyethylene (HDPE). PE is a semicrystalline material, with the density of the crystalline phase being approximately $1 \text{ g cm}^{-3}$ at room temperature, and that of the amorphous phase $0.854 \text{ g cm}^{-3}$. The presence of branches in the chain perturbs the formation of crystals. LDPE has the greatest amount of short and long chain branching and therefore has the

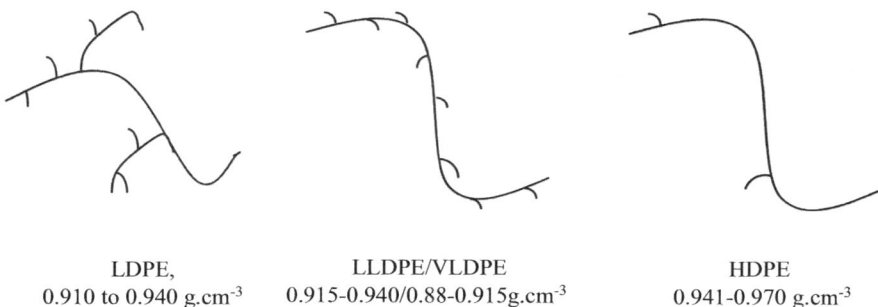

| LDPE, | LLDPE/VLDPE | HDPE |
|---|---|---|
| 0.910 to 0.940 g.cm$^{-3}$ | 0.915-0.940/0.88-0.915g.cm$^{-3}$ | 0.941-0.970 g.cm$^{-3}$ |

**Fig. 1** Polyethylene classification according to branching and density

lowest density. HDPE shows mostly linear structure, and thus a higher density. Although the density of LLDPE is similar to that of LDPE, it is made using a different chemistry, which leads to a different branching structure.

A simplified explanation for this is shown in Fig. 2. When free radical chemistry is used to make PE (Fig. 2a), the active centers are free radicals generated by decomposition of a chemical initiator such as a peroxide. In step 1 of Fig. 2a, the peroxide decomposes to give two highly active free radicals. The free radicals R* react with a monomer (step 2) to begin chain growth. Here, the radical is "transferred" to the monomer unit. This process of propagation continues (step 3) with the active center moving further and further away from the initiator fragment R that started the chain. This implies that the reactivity depends on the nature of the chain end rather than on the nature of the initial active center. Because polymerization conditions are such that the radicals are highly reactive and relatively mobile, the free radical can also be transferred to a growing (or dead) polymer chain, as in step 4. Thus, free radical chemistry leads to the formation of long chain branches and to many branches on a single macromolecule.

One can also encounter internal chain transfer, or backbiting (step 5), whereby the radical transfers backward on the growing chain, leading to the formation of short chain branches. This situation is very different from that encountered with a coordination catalyst (used to make HDPE and LLDPE), as shown in Fig. 2b. Here, the active center is a fixed metal atom (e.g., titanium) in the case of a Ziegler–Natta (ZN) catalyst. Polymerization begins when the first monomer unit coordinates with the active metal. Growth proceeds via an insertion mechanism, whereby the second monomer is inserted between the growing chain and the active center (step 1 in Fig. 2b), and continues in this way until the chain terminates and a new chain is formed on the same active center. This implies (1) that the active center participates in *every* insertion step and is extremely important in determining rates, selectivity, etc. (exactly the opposite of the situation in Fig. 2a); and (2) highly linear chains are favored and it is extremely difficult to form long chain branches. Thus, the only way to control the density of LLDPE is through the addition of alkenes, the linear tails of which form short chain branches in the polymer backbone.

PP is based upon propylene as monomer and its production represents one third of the global production of polyolefins. PP can only be made via a catalytic process;

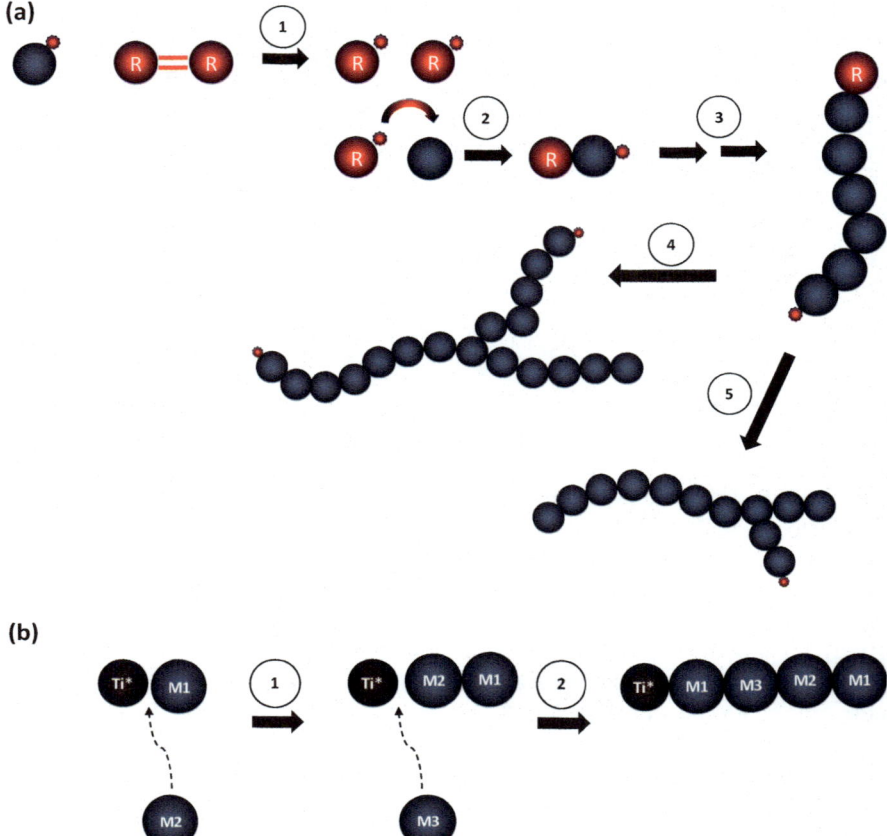

**Fig. 2** Different chemistry results in different structures. (**a**) The free-radical chemistry used for LDPE is monomer-centered, meaning that the active center is mobile and the growing chain end determines the rates at which the different steps occur. (**b**) When coordination catalysts are used, chain growth is by insertion between the active center (which participates in every step) and the growing chain

free radical polymerization is not feasible because propylene decomposes before the conditions for free radical formation are reached. The asymmetrical nature of propylene leads to different stereochemical microstructures (tacticity) for PP, as shown in Fig. 3. The orientation of the propylene methyl group in the resulting polymer significantly influences crystallization of the polymer and, thus, affects the physical and mechanical properties of PP. Catalysts used for PP production can be designed to control the stereospecificity of the growing chains (not the case for PE). Atactic polypropylene (*a*PP) is amorphous due to its random arrangement of methyl groups. Isotactic PP (*i*PP) and syndiotactic PP (*s*PP), are semicrystalline polymers with relatively high melting temperatures and high mechanical strength. Commercially, *i*PP holds the major market share, mainly because it is produced with

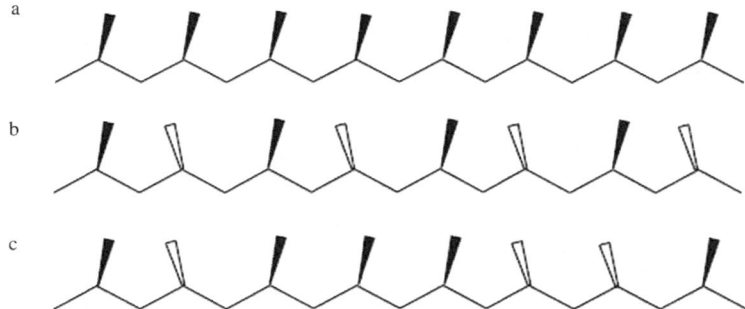

**Fig. 3** Main types of polypropylene: (**a**) isotactic, (**b**) syndiotactic, and (**c**) atactic [1]

(relatively) inexpensive heterogeneous ZN catalysts (and to a much lesser extent with metallocenes), whereas *s*PP which can only be produced with specific (post) metallocenes, which have lower activities and increase the cost of the process [1].

This very brief introduction to PE and PP presents a relatively simplistic view of what are very complex materials. For a more in-depth understanding, the reader is referred to several books on the subject [2–4].

# 3    Polyolefin Production Processes

Both the production process and the chemistry employed have a strong impact on the physical and chemical properties of polyolefins. Because no single process can economically produce all commercially important grades of polyolefins, there are various production processes in use. These processes can be divided into two broad families, based essentially on the chemistry used for the polymerization (and thus the pressures required for the process to function) and on the phase in which the reaction is carried out (as shown in Fig. 4). All these processes are continuous due to the very high production rates required for economic operation and the need to respond to the high demand for these polymers. In this section, we briefly describe some major points regarding the types of processes used for polyolefin production. For a detailed discussion, the reader pointed to chapter 4 of Soares and McKenna [5].

## 3.1    High Pressure Processes (Radical Polymerization)

Radical polymerization has been used to produce LDPE since its development by ICI in 1933. Tubular or autoclave reactor geometries are generally used with operating temperatures of 150–350°C and pressure varying between 1200 and 3500 bar. The combination of free-radical chemistry and harsh reaction conditions

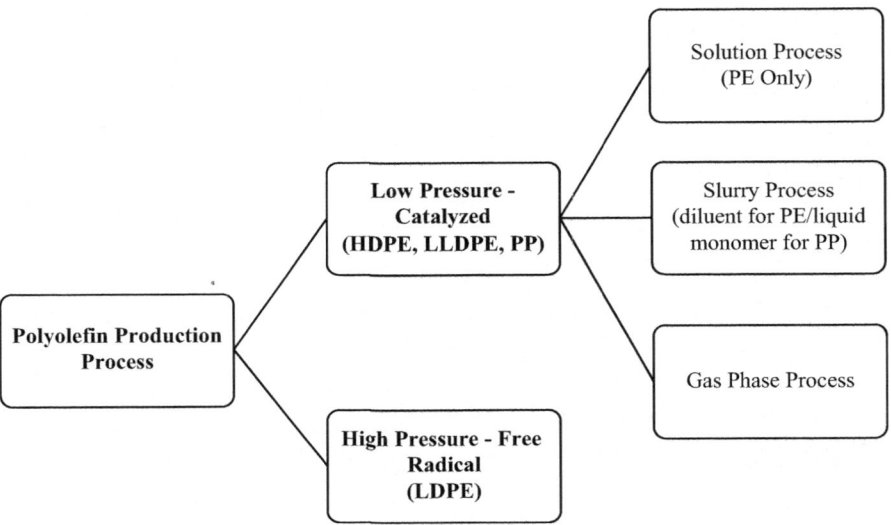

**Fig. 4** Classification of commercial polyolefin production processes

allows generation of the long chain branches that are characteristic of LDPE. In addition, free-radical polymerization allows incorporation of polar comonomers, which is impossible with commercially viable olefin polymerization catalysts. Thus, the high pressure process is the only way to make such materials.

Because autoclave reactors and tubular reactors have different residence time distributions, temperature profiles, and degrees of mixing, very different materials are produced from the different types of reactors.

## 3.2 Low Pressure Processes (Catalytic Polymerization)

Solution, slurry, and gas phase processes are different types of low pressure processes. About 80% of global polyolefin production is carried out using these processes because of their energy efficiency compared with high pressure processes, and the fine control they offer over certain aspects of polymer structure as a result of development of high activity olefin polymerization catalysts.

*Solution processes* are polymerizations carried out in the liquid phase, with reactor temperatures in the range of 140–250°C, so that the produced polyolefin remains dissolved in the reactor diluent. Autoclave and, to a lesser extent, tubular loop reactors are used for this purpose. Reactor volumes are in the range of 3–15 m$^3$ and typical residence times are 1–20 min. Because the reaction is carried out in the homogeneous phase and catalysts are unsupported and dissolved in the solvent, there is no mass transfer resistance. This, coupled with the high temperature needed to keep the polymer in solution means that the polymerizations are quite rapid.

This process is commonly used to produce different PE grades (e.g., ethylene/1-octene copolymers) with soluble metallocene and ZN catalysts. Ethylene-propylene-diene (EPDM) rubbers are also produced with this process, usually by employing soluble vanadium-based ZN or titanium-based postmetallocene catalysts. EPDM is polymerized at a lower temperature due to its amorphous nature. Molar masses of PEs produced in solution are generally low because of the high reactor temperature and difficulty in keeping high molecular weight (MW) molecules in solution [5]. The rest of this chapter deals with supported catalysts.

*Slurry processes* employ a heterogeneous catalyst, and the reactor can operate with two (liquid and solid) or three (gas, liquid, and solid) phases. Polymerization occurs inside solid supported catalyst/polymer particles, which are suspended in a suitable inert diluent. Diluents are of two types: (1) For PE processes, diluents are short chain alkanes, typically supercritical propane (Borstar process), isobutane, or *n*-hexane. (2) For PP processes, the diluent is usually liquid monomer. Reactors are generally followed by flashing units, where the absorbed diluent is desorbed from the polyolefin grade and recycled back into the process. Reactors of choice are loop reactors or, increasingly less frequent, stirred autoclaves. Typical reactor temperatures range between 75 and 100°C, and reactor pressures vary from about 8 to 65 bar, depending on the diluent, reactor configuration, and polyolefin grade being produced. Residence times of the slurry inside the reactor can vary from 45 min to 5 h, depending on reactor type and number of reactors.

The major advantage of slurry processes is their relatively favorable heat transfer conditions. All olefin polymerizations are highly exothermic, so the evacuation of energy from the reactor is crucial for safe operation and property control. Liquids are better fluids for heat transfer than gases; thus, the overall heat transfer coefficient in a slurry reactor is much higher than that in a gas phase process. The most common type of reactor is the loop reactor, which is essentially a very long tube, folded back on itself, with a very large recirculation pump to ensure an even flow of suspension inside the reactor. The linear velocity of the fluid inside the reactor is of the order of meters or tens of meters per second, which also helps with heat transfer. The loading of solids in a slurry loop reactor can have a significant impact on the viscosity of the suspension and, therefore, needs to be controlled precisely. The major limitation of slurry processes is that the product range is typically limited by the solubility of amorphous and/or low MW polymer in alkanes. The shorter the alkane used as diluent, the less this is a problem. However, it is not easy to make materials such as LLDPE (high amorphous content) or elastomers such as ethylene-propylene rubber in these processes because a significant fraction of the polymer dissolves in the continuous phase of the reactor and is then deposited on cold surfaces such as the reactor wall or the blades of the recirculating pump (referred to as fouling). It is therefore necessary to use gas phase processes in certain instances.

*Gas phase processes* are also multiphase processes. Polymer is formed on heterogeneous catalysts and suspended (or mixed) in the presence of a gaseous stream that contains inert(s), hydrogen, short chain alkanes, and (co)-monomer(s).

In a fluidized bed reactor (FBR), the catalyst/polymer particles remain suspended in a fluidizing gas phase that is fed into the bottom of the cylindrical reactor. FBRs are the only economically viable means of making PE in the gas phase, because the relative velocities of gas particles are high enough that sufficient heat can be removed from the reactor. In addition to FBRs, mechanically stirred powder bed reactors (i.e., not fluidized; particles are agitated as the gaseous stream flows over them) can also be used to make PP. The reason for this is that PP processes generate less heat per cubic meter than PE processes, and PP melts at 30–40 K higher than PE. A fourth reactor type, the Spherizone riser-downer reactor from LyondellBasell is also used. Gas phase reactors generally operate in the temperature range of 70–110°C, and under pressures of 20–25 bar, depending on process technology. Residence time of the particles inside one reactor can be from 1.5 to 3 h. Reactor operation can be either in dry mode (no liquefied inert condensing agent is used) or condensed mode (a liquefied inert condensing agent is used to improve heat transfer inside the reactor). Reactor volumes can vary from 50 to 150 m$^3$ [5]. Companies that have their own gas phase licensed technologies include Univation Technologies, LyondellBasell, INEOS, and Mitsui.

Heat transfer in gas phase processes is a challenge. Economics pushes producers to make the polymer as cheaply as possible, but it is harder to remove the heat from a gas phase process than from a slurry process. Certain improvements, such as the use of alkanes in the feed to the reactor, can help to alleviate this problem. Alkanes such as iso-pentane can significantly increase the heat capacity of the gas phase, allowing it to absorb more heat. Furthermore, if the temperature of the feed stream drops below the dew point of the mixture, the latent heat of vaporization of the liquid droplets fed into the reactor can also help to control reactor temperature (at the cost of increasing the load on the downstream purification units).

Despite these challenges, gas phase processes remain the only means of making polyolefins containing significant amounts of amorphous material. The vast majority of LLDPE and elastomeric copolymers are made in this way. Thus, it is possible to make any type of polyolefin using the gas phase process. In addition, gas phase processes have lower capital and operating costs than slurry processes. The choice of which process to use depends on the markets the production line is to serve.

*Mixed phase processes* can also be found at commercial production sites. Processes such as the Borstar process from Borealis or the Spheripol process from LyondellBasell (there are others!) employ a combination of slurry loop (s) preceding one or more FBRs. The idea here is that one can benefit from very high reaction rates at the front end of the process to make HDPE (e.g., Borstar, Hypol from Mitsui) or isotactic PP (e.g., Borstar or Spheripol), and then send the still active polymer/catalyst particles to gas phase reactors to produce more amorphous or low MW materials. This approach gives far more intimate mixing of different polymer phases than coextruding two different materials made in different production lines.

Note that this is a very rapid overview of a broad and complex topic, intended simply to help the reader understand the more detailed discussion of catalysts, and metallocene catalysts in particular, that follows. Furthermore, the rest of this chapter focuses on PE made in slurry and gas phase processes; solution PE and PP are not considered. The solution process uses molecular catalysts and, in many ways, the physical challenges of making supported catalysts for PE are similar to those encountered for PP. This does not mean that the PE and PP catalysts are the same – in fact, the chemical natures of these catalysts are quite different, not least because of the need to modify PP catalysts to obtain good control over the stereospecificity of the polymers. However, issues such as particle size/structure and leaching can be viewed as being similar for both.

# 4    Supported Olefin Polymerization Catalysts

Since their commercialization in the 1950s, transition metal-based catalysts have been continually modernized such that they provide better and better molecular control of the polymerization process, and are faster and faster. Although it might be disappointing for engineers to admit, it is advances in catalyst chemistry, not technology, that are the main reason for tremendous growth in the polyolefin industry, as well as for the evolution of process technology. Figure 5 shows that,

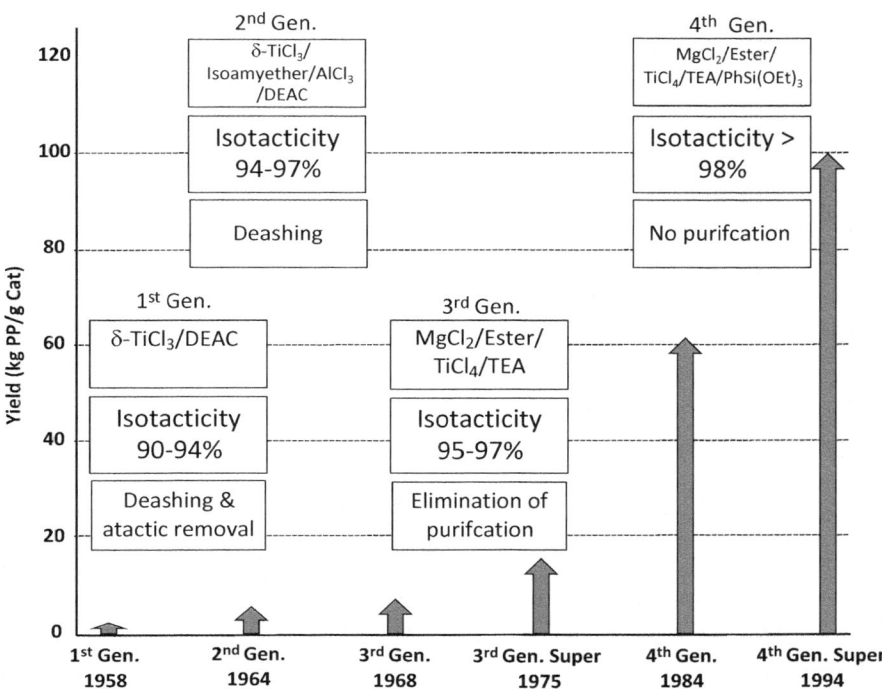

**Fig. 5**  Developments in polypropylene (PP) Ziegler–Natta catalysts. Adapted from Galli et al. [6]

between 1958 and 2001, the average productivity of a PP catalyst increased 100-fold, meaning that engineers had to figure out how to get 100 times more monomer into the process while removing 100 times more heat! Since 2001, we have seen the development of increasingly sophisticated catalysts entering commercial exploitation, particularly metallocene catalysts (see Sect. 4.3).

It is useful to take a quick look at the major families of commercial catalysts. The discussion is limited to heterogeneous, or supported, catalysts. Heterogeneous (i.e., supported) forms of these catalysts are employed in the low-pressure slurry and gas phase industrial processes. Ideally, the material used to support the catalyst should have some surface functionality and reasonable porosity, friability, and particle size. Surface functionality of the support allows it to help retain the active site(s), but physical properties (porosity, friability, particle size, etc.) play a crucial role in active site(s) distribution inside the catalyst particles. (Co)monomer (s) diffusion during polymerization can further impact the polymer molar mass distribution, comonomer composition distribution, and final product morphology (as discussed in Sect. 6).

All the supported catalysts discussed here have one thing in common – a support! Although other options are possible, the most commonly used supports for olefin polymerization catalysts are silica gel and magnesium dichloride. The support can serve at least two purposes: (1) a vehicle for bringing the active sites into the reaction zone and creating a solid polymer particle, and (2) helping to modify the activity/productivity of the catalyst if the support is treated in an appropriate way [7–11]. Thus, supports should have high surface areas to carry a sufficient number of active sites, decent porosity to facilitate mass transfer of the reactants to the active sites, and an appropriate balance of mechanical properties to ensure that catalyst particles are strong enough to be transported to the reactor, but friable enough that once the reaction begins they will fragment to produce one polymer particle per catalyst particle (see below). This polymer particle should have a regular shape, high bulk density, and sufficient porosity for the polymerization to continue until product is removed from the reactor.

In general, inorganic oxide support materials have two structural levels, microparticles and macroparticles. Nanosized microparticles (or micrograins) combine to form bigger particles of diameters typically in the range of 10–100 µm, which are usually termed macroparticles (or macrograins). These macroparticles are highly porous, with surface areas of the order of several hundred square meters per gram of support. After impregnation with active sites and/or cocatalysts, the active species are assumed to be evenly distributed throughout the macroparticles. As soon as these supported catalyst particles are injected into the reactor, polymer starts to accumulate within the pores of the macroparticles as a result of polymerization at the active sites. At this point, such polymerizing catalyst particles are generally referred to as growing polymer particles. Accumulation of polymer inside pores generates hydraulic stress and, at a certain point, when the physical bonds holding the micrograins are unable to bear the hydraulic stress generated by the formed polymer, the so-called phenomenon of particle fragmentation starts.

The concept of fragmentation is very important, and significant effort has been dedicated to understanding how it occurs, the role of the support, etc. Particle fragmentation gives access to active sites that are buried under the polymer formed in the particle. Note that, in the case of $MgCl_2$-supported ZN catalysts, the fragmentation step can expose new active sites that are integrated into the support material. However, in the case of silica-supported metallocenes, the active sites are generally located on the surface. By the end of this fragmentation step, the original biphasic support material (i.e., the pore space and solid particle) is converted into a triphasic mixture containing catalyst-impregnated solid fragments of the support material, continuous polymer phase embedding these catalyst-impregnated solid fragments, and the porous space through which the reactants are transported. Ideally, one catalyst particle should generate one polymer particle, which continues to grow by expansion as a result of polymer formation at the active sites.

Different mechanisms for this critical step of particle fragmentation have been proposed, but the most widely accepted for silica-supported catalyst particles is that the polymer layer is first formed on the outermost easily accessible particle surface, leading to higher inward stresses due to polymer formation and, therefore, fragmentation proceeds from outside toward the center of the particle. Typically, the time span of fragmentation completion is several tens of seconds for silica-supported catalysts [12, 13]. If the support material is weak, uncontrolled fragmentation can generate fine polymer particles (i.e., diameters below 200 μm), which are detrimental for industrial reactors. On the other hand, strong support material can lead to pores becoming blocked with polymer and, consequently, little or no catalytic activity. Therefore, one has to select a support material carefully so that both problems can be avoided. Nevertheless, adjustment of reactor conditions and addition of certain comonomers and inert condensing agents to the reactor are helpful in controlling the fragmentation step, especially during gas phase processes where heat transfer control is generally problematic [13].

All support materials have different properties, including matrix strength or fragility, leading to differences in the fragmentation step. Therefore, the supported metallocenes behave differently under similar reaction conditions.

The fragmentation step has been extensively studied, so the interested reader is referred to a number of references for detailed discussions [13–23]. The exact mechanisms underlying this process are still under discussion, but certain points are worth retaining:

- Fragmentation is rapid relative to the residence time of commercial reactors (seconds or tens of seconds versus several hours).
- Not only the mechanical properties of the support are important in determining the outcome of this step, but the rate of polymerization and the mechanical properties of the polymer also play a role [24, 25].
- $MgCl_2$ typically fragments faster than silica.
- Poorly controlled fragmentation leads to significant operating problems in the reactor.

It should not be forgotten that different kinds of catalysts behave in different ways and interact differently with different supports. Thus, the active site–support complex plays a significant role in determining the final polymer properties. However, when considering the use of supported catalysts, challenges such as heat and mass transfer inside the particles as they grow must be kept in mind. Clearly, the support type and the resulting polymer particles influence the characteristic length and time scales for these phenomena, and also play a role in the outcome of the polymerization. Before moving on to the main topic of this chapter, which is the effect of the support properties on the reaction rate and properties of ethylene polymerization on supported metallocenes, we take a brief look at the three major families of catalysts in widespread commercial use.

## 4.1   Ziegler–Natta Catalysts

Ziegler–Natta (ZN) catalysts are formed by the interaction of main group metal alkyls with halides or other derivatives of transitions metals of groups 4–8 of the periodic table. Both homogeneous and heterogeneous forms of ZN catalysts can be found in commercial applications, but we will focus on supported catalysts [26, 27]. In the metal alkyl component of the ZN catalysts, the metal atom belongs to group 1–3 of the periodic table [27, 28]. The metal alkyl component is also known as the activator or cocatalyst, whereas the transition metal halide or other derivative part is mostly referred to as the catalyst or, perhaps more accurately, as the precatalyst because alone it is inactive in olefin polymerization. In this chapter, we will refer to the latter as catalyst. Interaction of catalyst and cocatalyst enables the generation of species that are active for olefin polymerization and generally referred to as active sites. The cocatalyst alkylates and reduces the transition metal center of the catalyst to produce the active site and, therefore, active site generation is considered to be a two-step process – it is therefore important that the cocatalyst be uniformly distributed throughout the support particle. Trimethylaluminum (TMA), triethylaluminum (TEA), triisobutylaluminum (TIBA), trioctylaluminum (ToA) and diethylaluminum chloride (DAEC) are some of the preferred cocatalysts for ZN catalysts. It is important to mention here that the term "Ziegler catalyst" is sometimes used in the literature when these catalysts are employed in ethylene polymerization, whereas "Ziegler–Natta catalyst" points to their use in propylene polymerization.

Supported (or heterogeneous) ZN catalysts are used commercially in slurry and gas phase processes. These catalysts have dominated the polyolefin industry for the last 50 years due to their high productivity and relatively low production cost (on the order of tens of euros per kilogram of catalyst). $TiCl_4$ supported on $MgCl_2$ is the most general form of heterogeneous catalyst for polyolefin production. There are various synthesis routes for these catalysts to guarantee high activity, good molar mass control, good comonomer incorporation, stereoselectivity, and polymer morphology [29]. $TiCl_4/MgCl_2$ can also be supported on silica gels if the titanium–magnesium salt complex is deposited as a thin layer on the surface of the supports.

ZN catalysts are generally divided into different generations. Soares [27] divided them into four generations (see Fig. 5), Chadwick et al. [29] classified them into five generations, and Galli and Vecellio [6] defined four generations and two subgenerations. Without going into too much detail, ZN catalysts of the earlier generations were based on $TiCl_3$ catalysts activated with DAEC and $TiCl_3$ modified with donors to enhance stereoselectivity. Later generations are based on $TiCl_4$ supported on $MgCl_2$. Further improvements include chemical modification of the support, better internal and external donors, and better control of morphology to help reduce or eliminate the need for palletization. Each generation shows gains in terms of property control and productivity. With current ZN catalyst systems, it is possible to make tens, or even hundreds, of kilograms of polymer per gram of catalyst (with high stereoregularity in the case of PP catalysts).

Like the Phillips catalysts discussed in Sect. 4.2, supported ZN catalysts are referred to as "multisite catalysts." Describing them as $TiCl_4$ supported on $MgCl_2$ is perhaps overstating the complexity of these catalysts. It is thought that possible interactions between Ti atoms, alkylating cocatalysts, and $MgCl_2$ support create different environments for different Ti active centers. This means that neighboring Ti atoms can react at different rates, make chains of different length, show different hydrogen sensitivity, and incorporate comonomers at different rates (generally, the longer the polymer chains made on an active ZN site, the less they seem to be able to incorporate comonomers). Given that each "type" of site produces a polymer with a polydispersity index of 2, the fact that we have several sites producing different polymers with different average MWs means that the overall MWD of polymer produced using a ZN catalyst is fairly broad, with an overall polydispersity (typically) between 4 and 8.

## 4.2 Phillips or Chromium Catalyst

Phillips or chromium catalysts were discovered by J. Paul Hogan and Robert L. Banks at the Philips Petroleum company in the last half of 1951 [30], more or less when Karl Ziegler disseminated his results on Ziegler catalysts. A common chromium catalyst is made by supporting Cr(III) hydroxide on silica or aluminosilicate, followed by calcination in dry air at high temperatures. Chromium catalysts are generally classified into two main families: (1) those based on chromium oxide, known as Phillips type; and (2) those based on organochromium compounds. They are different from ZN catalysts in the following respects:

- They do not exist as molecular catalysts, only supported on calcined silica gel supports.
- They do not need an aluminum alkyl cocatalyst to form the required metal–carbon bond.

- Polymerization activity, MWD of the polymer, and long chain branching inside the polymer are significantly influenced by the calcination temperature and calcination procedure.
- Hydrogen is not an effective chain transfer agent for this catalyst.
- Long induction periods are very common during polymerizations with this type of catalyst [27, 29].

These catalysts are first pre-activated by calcination at temperatures of 200–900°C. During thermal activation, the Cr species attaches to silica by reaction with the surface silanols at 200–300°C. Further rise in temperature (i.e., >500°C) removes the neighboring silanols [27]. During this activation step, Cr(III) is converted into Cr(VI), which is not active in olefin polymerization and must be further reduced to Cr(II) to be able to polymerize ethylene. The transformation of Cr(VI) to Cr(II) takes place inside the reactor when Cr(VI) comes into contact with ethylene monomer. This is the typical route applied in commercial plants using this catalyst. The complexity of the activation step means that chromium catalysts are also multisite catalysts, with a variety of active centers that can behave in very different ways. The polydispersity index can be extremely high for a Phillips catalysts, approaching 10, with McDaniel [30] reporting values as high as 65. An exhaustive review describing in detail the specific features of Phillips catalysts, reaction mechanism, and polymer properties can be found in the works by McDaniel [30, 31].

## 4.3 Metallocene Catalysts

Metallocene catalysts consist of a transition metal atom sandwiched between two cyclopentadienyl (Cp) or Cp-derivative rings, as depicted in Fig. 6 [27]. The structure of olefin polymers was uncovered by Ernst O. Fischer and Geoffrey Wilkinson in 1952, for which both were awarded the 1973 Nobel Prize [32]. The transition metal atom usually belongs to group 4 and is mostly zirconium (leading to zirconocenes), titanium (leading to titanocenes), or hafnium (leading to hafnocenes). Metallocenes are soluble in hydrocarbons and show only one type of active site upon activation. Commercially significant activities from metallocenes were possible after the discovery by Sinn and Kaminsky of methylaluminoxane (MAO) as an efficient activator (giving about 10,000 times higher metallocene activity than when activated with $AlXR_2$) [33–35], almost three decades after the results reported by Breslow and Newburg [36] and Natta et al. [37]. The discovery of MAO re-ignited scientific research in the field of metallocene catalysis. Of the different metallocene activators or cocatalysts (aluminum alkyls, borates, fluoroarylalanes, trityl and ammonium borate, aluminate salts, etc. [38]), MAO and its different modified forms seem to be the most widely used, both in research and production. MAO consists of alternate arrangements of aluminum and oxygen atoms, the free valances being saturated with methyl substituents. Although

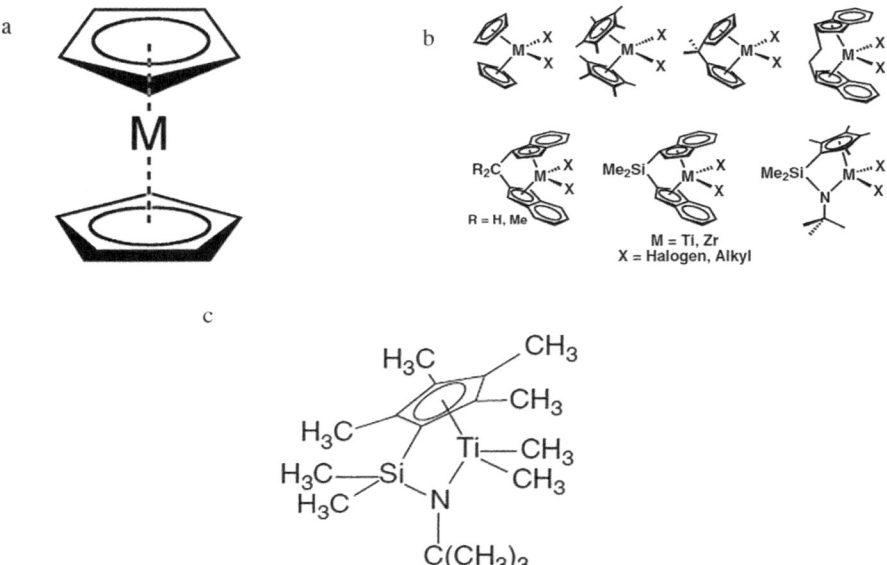

**Fig. 6** Schematic presentation of metallocene catalysts (**a, b**) and CGC (**c**)

metallocenes are used to a lesser extent than ZN catalysts, in 2010 over 5 million tons of polyolefins, especially different grades of PE, were produced commercially using MAO as an activator, pointing to the fact that metallocenes are still important [34].

The reactivity of these catalysts toward olefins can be tailored by variations in the electronic and steric environment around the transition metal, which has enabled the production of polyolefins with reasonably well-controlled molar mass distribution (ideally with a Flory distribution, having polydispersity of 2) and polymer microstructure (comonomer content and distribution, short and long chain branching, and polymer tacticity) [32]. The molar masses of the PEs produced with metallocene catalysts can vary over a wide range of 18,000 to 1.5 million g mol$^{-1}$. Reactor temperature, metallocene-to-ethylene ratio, and amount of hydrogen injected into the reactor can be used as molar mass control handles [39]. Various classes of metallocenes have now been developed, including *ansa*-metallocenes in which the two Cp rings are connected through bridges of different types, which allows modification of the ligand–metal–ligand angle (commonly known as the bite angle) [27].

Some homogeneous or soluble metallocene/MAO catalysts have found industrial applications in solution processes (such as Dow ELITE, Nova Surpass, Exxon EXACT) [40] where the operating conditions are such that the whole reaction mixture is a monophase liquid [41]. However, the high cost and quantity of cocatalyst required, difficulty in injecting the catalyst into the existing slurry and gas phase reactors, low processability, and poor polymer morphology have inhibited their direct use in existing slurry and gas phase processes. Nevertheless, there has been progress in recent years on the support or immobilization of metallocene/MAO catalysts on suitable carriers (or supports), such as silica, alumina, and alumina-silica. This has overcome most of the problems cited above, allowing this type of catalyst to be used as "drop-in-technology"; that is, as new catalysts that can be used in an existing polyolefin plant without any significant structural changes [38, 41, 42] (see Sect. 5).

As mentioned above, metallocene catalysts are generally known as "single-site" catalysts because all the active sites are thought to behave in the same way. This is in contrast to the two multisite catalysts mentioned above (i.e., ZN and chromium catalysts), and is part of the overall attractiveness of metallocenes in their supported form. Certain disadvantages are associated with production of very narrow, precise polymer MWDs, such as difficulty in processing metallocene-made PE (the low MW tail obtained with multisite catalysts acts as a "lubricant" in the extrusion process), but precise control over the MWD and the incorporation of comonomers make these catalysts useful in terms of tailoring polymer properties [27, 41]. Some of the challenges can be overcome by using multireactor technology. Using more than one reactor allows the process to include zones with different concentrations of monomer and hydrogen and, thus, produce different macromolecules on the same sites in a controlled manner. Also, the use of tandem catalysts (i.e., supports carrying more than one type of catalyst) with components of varying comonomer incorporation abilities, stereoselectivities, hydrogen responses, and chain walking abilities in one reactor is becoming popular. The interested reader can find a detailed discussion of the industrial use of multisite metallocenes catalysts in a review by Stürzel et al. [43].

# 5  Supporting Metallocene Catalysts

The use of drop-in technology (i.e., supporting metallocenes) has enabled the use of metallocene catalysts in a wide range of processes. However, how active sites are supported is important for use in gas or slurry phase processes. There is a range of different types of supports for metallocenes [40, 41, 44–48], but the most widely used seems to be silica because of its low cost, ease of handling, and good reactivity toward the metallocenes and cocatalysts due to the presence of various hydroxyl groups on its surface and interior.

Despite recent advances in supporting metallocenes, heterogeneous or supported metallocene/MAO catalysts are not as active as their homogeneous analogs, perhaps due to different side reactions that probably depend on the method of supporting. In addition, leaching of the supported catalyst can cause reactor fouling, which is another operational problem. In a nutshell, the performance of a silica-supported metallocene can be linked to the various factors shown in Fig. 7. The message of Fig. 7 also includes the often ignored fact that, while studying the role of physical properties of supported metallocenes in polymerization kinetics, etc., the

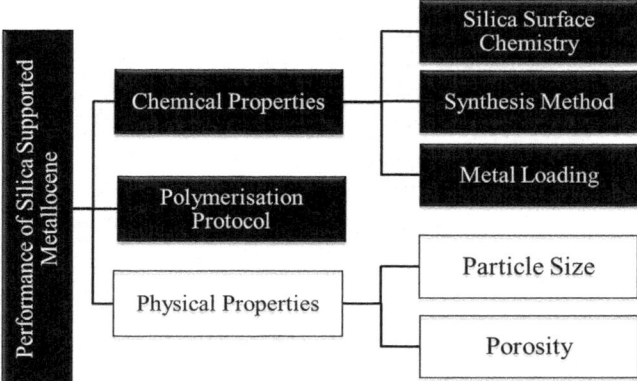

**Fig. 7** Overview of the factors affecting the performance of a silica-supported metallocene catalyst in olefin polymerization

chemical properties and polymerization protocol (i.e., how ingredients are added into the reactor) must be kept as constant as possible, and (at least as important) the physical properties should never be ignored in "chemical" studies!

Continued industrial and academic research has partially mitigated these issues by providing better understanding about the interactions of catalysts and/or cocatalysts with various supports. This has led to wider use of metallocenes in commercial processes, either as an individual catalyst or as a component of multisite supported catalysts [41–43, 49]. Despite this progress, there remains a number of important points that need to be investigated in terms of improving the supporting process, understanding how the support influences the performance of the catalyst, and, in particular, how the characteristics of the support influence the polymerization.

Before focusing on the impact of the support morphology on metallocene preparation and activity, we briefly describe the methods used to support metallocenes on silica gel particles. Readers interested in more details of site structure and types, and current methods used to support active sites should see the works of Collins et al. [50], McKnight and Waymouth [51], Theopold et al. [52, 53], Severn et al. [41], McDaniel [30, 31], and Hlatky [46].

## 5.1    Silica as a Support

Put simply, a heterogeneous or supported metallocene is created by putting catalyst into the pores, and then onto the surface, of a solid particle. However, which support to use and how to "put" the sites into and onto the particle surface is a complex process. One needs to consider not only the type of metallocene(s) to use, but also the type of support, the surface chemistry of the support, the means of putting site and support into contact, and the steps needed to ensure that the site remains inside the particles. It is easy to imagine that different supports offer different advantages and disadvantages for supporting metallocenes. It has been 40 years since the disclosure of US patent 4161462 in 1976, in which the inventors showed that 1,2-polybutadiene can be used as a support for $Cp_2TiCl_2$. Ever since, different materials have been used for supporting metallocenes, including inorganic oxides such as silica and alumina, zeolites (which are aluminosilicates), mesoporous silicates, magnesium dichloride, clays, layered double hydroxides, and polymers [40, 41, 44–48]. However, the most widely used support seems to be silica due to its low cost, ease of handling and good reactivity toward the metallocences and cocatalysts due to the presence of various hydroxyl groups on its surface and interior. The rest of this chapter focuses on silica supports.

The use of silica as a support in olefin polymerization dates back to the 1950s and remains the most popular material for catalyst supports in the polyolefin industry. Although silica and other supports are often referred to as being inert, this is not strictly speaking true. Silica has functional groups on its surface, and these show reactivity toward the catalysts/cocatalysts and can alter the nature of the active sites. Although silica is chemically inert in terms of the polymerization steps, the nature of the surface is important in determining the behavior of the active sites. Interactions between silica and the components of the active sites can have a significant influence on productivity, comonomer incorporation, and stereoselectivity of the supported metallocenes, as well as on the molar mass distribution of the produced polyolefins. The physical properties of the silica support, which can be altered to varying degrees during silica synthesis, can also play a vital role in determining the final performance of metallocenes and/or cocatalysts supported on silica in olefin polymerization. Little systematic work has been done to understand the impact of the geometric factors of the supports on the polymerization process and the resulting polymers made with metallocenes.

The following properties of silica supports play crucial roles in catalysis by heterogenous metallocenes:

- Chemical properties and surface chemistry (number and type of surface species such as silanol, silyl-ether, and Lewis and/or Brønsted acid sites) [41, 54–60]
- Mechanical properties of silica (i.e., friability, which refers to the strength of the silica matrix against internal stresses generated by, for example, polymer accumulation during polymerization and attrition) [61]
- Physical properties (e.g., particle size, pore volume, pore size, pore size distribution, surface area)
- External and internal pore structure (i.e., the method of silica manufacture)

In terms of surface chemistry, the surface of amorphous silica gel is saturated in silanol groups in fully hydroxylated and unmodified forms. Three distinct types of silanol groups (SiOH) are present on the silica surface along with siloxane groups (–Si–O–Si–). In addition to silanols and oxygen-bound siloxanes, water is also structurally bound inside the silica skeleton and in very fine ultramicropores with diameters of less than 1 nm [56]. Calcination or dehydroxylation is required to remove adsorbed water and most of the silanol groups because they are poisons for metallocenes. Calcination refers to the thermal treatment whereby the support material is fixed or fluidized in an oven, multiple hearth furnaces, or rotary oven. The three steps of heating, calcination, and cooling have their own distinct time and temperature ramps, hold times, temperatures, and optional agitation. All of these parameters are set in such a way that particle sintering is avoided during the whole process [54]. Dehydroxylation is generally done under vacuum, without any inert or air flow, in two steps (heating and cooling). The duration and levels of the temperature ramps and hold times can also differ.

The final hydroxyl group density depends on the temperature and time of the thermal treatment, but usually stays between 1 and 5 OH $nm^{-2}$. It is important to mention that calcination can also alter the pore volume and pore size distribution; for example, increased calcination temperatures have been reported to decrease the pore volume and surface area of the support. Furthermore, one can also modify the silica surface with different compounds such as chloro-silanes, alkoxy-silanes, or disilazanes for specific applications in metallocene heterogenization [54].

At the risk of oversimplifying, the main aim of the different surface treatment methods is to generate an immobilization surface that does not poison the metallocene or the cocatalyst. In the case of unmodified heat-treated silicas, the hydroxyls and siloxanes act as fixation sites for metallocenes or the cocatalysts. However, there is no single optimal value for the concentration of these functionalities because different metallocenes and cocatalysts have different sensitivities. For example, Fig. 8 shows that the dehydroxylation temperature can have a strong influence on productivity, and that this influence is different for different types of active sites. By studying the most widely used silica dehydroxylation temperature range of 200–600°C, we showed that siloxane bonds can be formed in silica by dehydroxylating at a temperature $\geq$450°C (previously, it was believed that siloxane bonds are formed on silica surfaces at $\geq$600°C) [63]. Furthermore, we demonstrated that the most active $(n\text{-BuCp})_2ZrCl_2$/MAO catalyst was formed when the silica support was dehydroxylated at 600°C (considering the studied temperature range). On the other hand, for the $rac\text{-Et(Ind)}_2ZrCl_2$/MAO catalyst (where Ind stands for indenyl), the influence of temperature was quite different, and higher activities were observed at a silica dehydroxylation temperature of 200°C using a similar catalyst synthesis method but different metallocene. Other authors [59] have also reported similar results for PP polymerization using silica-supported $rac\text{-Et}$ $(Ind)_2ZrCl_2$/MAO catalyst prepared in a similar way to that used by Bashir et al. [62, 63].

The mechanical properties of the silica are important, but much more from the point of view of controlling fragmentation and morphology than of influencing the chemical nature of the catalyst itself, so the latter is not discussed here. However, other physical properties of the supports, specifically structural aspects such as pore size, pore size distribution, and surface area, are crucial in determining how the catalyst behaves and how the polymerization proceeds because they impact the distribution of the catalyst and cocatalyst throughout the solid particles. This is important during catalyst synthesis, (co)monomer(s) diffusion during

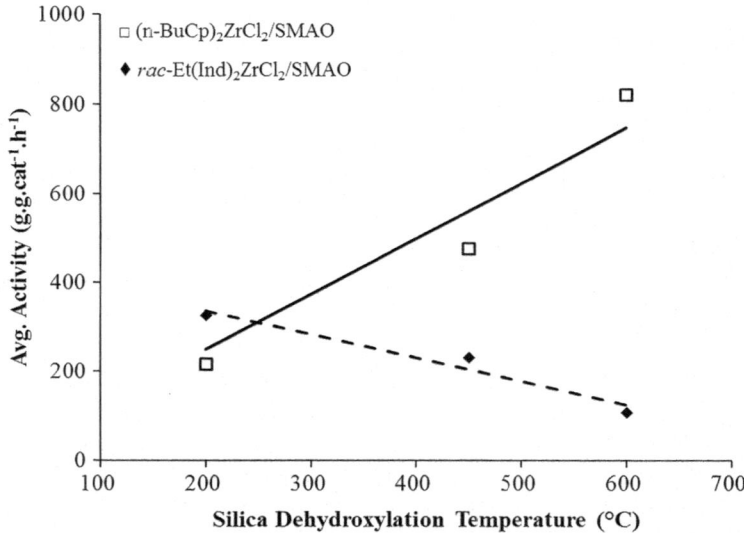

**Fig. 8** Effect of silica dehydroxylation temperature on the productivity of SMAO-supported ($n$-BuCp)$_2$ZrCl$_2$ and $rac$-Et(Ind)$_2$ZrCl$_2$/SMAO. Slurry phase polymerizations at 80°C, 8 bar ethylene pressure, TEA (2 mmol L$^{-1}$) as scavenger in $n$-heptane. Grace 948 silica was used as support. Catalysts were prepared and polymerized as discussed by Bashir et al. [62, 63]

polymerization, growth of molar mass of the polyolefin and, last but not least, during crystallization of nascent polyolefin chains within the porous support. Despite their significance, the impact of the physical properties of silica supports on the performance of supported metallocenes during ethylene (co)polymerization processes (i.e., slurry, gas, or bulk processes) is not as well explored as the impact of the chemical or surface properties of silica supports on the reaction kinetics of supported metallocenes in olefin polymerization. We return to this very important point in Sect. 6, but first we need to look at how the catalytic sites are put in/on the support.

## 5.2  Supporting Metallocene Precursors

Once a support (here silica) has been selected and treated, the metallocene and/or cocatalyst must be put into the pores and onto the surface. Several methods have been described in the open literature for preparing heterogeneous or supported metallocene catalysts. Here, we use the term "supported metallocenes" instead of "heterogeneous metallocenes." Each method involves a certain level of complexity and provides supported metallocenes with specific advantages. Generally, most methods used for supporting metallocenes fall into the following two broad categories [55, 64, 65]:

1. Physical adsorption supporting methods:

   (a) Support/metallocene/co-catalyst
   (b) Support/cocatalyst/metallocene
   (c) (Metallocene + cocatalyst)/support

2. Chemical tethering methods:

   (d) Support/functionalized metallocene/cocatalyst
   (e) Metallocene generation on the support

Physical absorption leads to supported metallocenes, where the bonding between the support and catalyst or cocatalyst is not very strong. Although some of these methods have found successful industrial implementation (usually in the gas phase) and provide supported metallocenes with commercially acceptable activities, selectivities, etc., the inherent problems of low activity compared with the homogeneous analog, catalyst leaching, multisite behavior, etc., are still being investigated and improved. When these methods are preferred, it is probably because of their simplicity and resultant low production costs, rather than the efficiency of tethering the site to the support.

In the support/metallocene/cocatalyst method (method 1a), the metallocene dissolved in a suitable solvent (e.g., toluene) is brought into contact with the silica. The hydroxyl or siloxane groups available on the silica react with the metallocene and fix it either coordinatively or covalently. Cocatalyst (e.g., MAO dissolved in toluene) is then added and coordinatively attaches to the supported metallocene, generating the active species. Washing steps are commonly applied after metallocene fixation and cocatalyst impregnation. The actual number of washing steps, the volume and type of hydrocarbon used, and the preparation temperature vary significantly. This method of supported metallocene synthesis is not preferred because the local steric environment of the metallocenes is influenced by close contact with the support surface, so there are very few examples of success using this method [54]. Furthermore, the formation of bidentate species during catalyst synthesis can significantly reduce the activity of such supported metallocenes in olefin polymerization [41, 58, 66, 67].

In the support/cocatalyst/metallocene method (method 1b), silica is first impregnated with cocatalyst, followed by washing and drying steps [68–74]. In the second step, the support impregnated with cocatalyst is suspended in a hydrocarbon to form a slurry, and a toluene solution of metallocene is added. Washing and drying steps are applied again and the final product is the supported catalyst. The reaction and drying temperatures, contact time for each step, and number of washing steps can all vary. The functional groups of silica act as fixation sites for the cocatalyst (e.g., TMA in MAO), whereas the absorbed MAO coordinates with the metallocene to form the active species in the second step. MAO supported on silica (SMAO) is also available commercially and one can directly support metallocene on such a commercial SMAO. A major benefit of this method is the avoidance of metallocene decomposition or deactivation by direct interaction with the functional groups of the support surface. SMAO is also suitable in cases where precontact between the metallocene and cocatalyst leads to over-reduction or deactivation of the catalyst [54]. It has also been shown that the efficiency of the final supported catalyst can be improved by orders of magnitude by correctly choosing the heat treatment during MAO impregnation of the silica support, duration of contact between the support and cocatalyst, and chemical modification of the silica support before and after immobilization [41, 54, 55, 57, 64, 75–79]. For these reasons, this method of supporting metallocenes has become one of the most widely used academically and industrially.

The (metallocene + cocatalyst)/support method (method 1c) involves mixing a metallocene and cocatalyst (usually MAO) in a suitable solvent prior to their addition on the silica support. This procedure has become one of the most frequently utilized methods of preparing supported metallocenes because it has a limited number of time-consuming steps, uses less solvent than most methods, and generates fewer byproducts. This approach can therefore reduce the production cost of supported metallocenes. Furthermore, in instances where the combination of catalyst and cocatalyst permits it, dissolving the two species in one pot can lead to better activation of some metallocenes because there are no diffusion limitations and fewer byproducts that can interfere with catalyst activity in solution. However, some complexes can deactivate upon precontact with cocatalyst, so the scope of this method is limited [54, 65]. The major problem with this method is that the mechanism of fixation of the metallocene + MAO mixture onto the silica surface is unclear.

An important advance in supported catalyst synthesis was the development of the "incipient wetness method," which allows commercial plants to save production costs by reducing the amount of solvents used and byproducts produced [54, 80–82]. In this technique, the pores of the support are filled in a controlled manner with solvent containing either metallocene/cocatalyst mixture or MAO alone. Alternatively, a solution of metallocene can be fed to the cocatalyst/support (i.e., method 1b). The total volume of the solution of active ingredients is typically 100–150% of the pore volume of the bare support (although occasionally the solution volume can exceed 150% of the total pore volume in order to shift from mud-point to slurry state) [72]. Capillary forces draw the solution into the pores of the support, leading (in principle) to uniform dispersion of the active ingredients throughout the porous particles [54, 80–82].

It is important to mention here that the solvents used during supported catalyst synthesis cannot be completely removed, even after vacuum drying the final catalyst at different temperatures, inert gas flow rates, etc. The amount of residual toluene, for example, can vary from ~1 to ~30 wt% of the final catalyst depending upon the drying conditions employed [54].

Silica-supported metallocenes prepared by the physical adsorption methods described in Sect. 5.2 can be prone to leaching; in other words, the extraction or desorption of the metallocene, cocatalyst, or metallocene/cocatalyst species from the silica support. Desorbed metallocenes are generally soluble in the reactor diluent and can polymerize in that phase in the presence of separately added cocatalyst (if the metallocene and cocatalyst together have desorbed, a separately added cocatalyst is not needed). This phenomenon is undesirable because it causes reactor fouling and, consequently, heat transfer problems. Clearly, it is more of a concern in slurry phase reactors than in gas phase olefin polymerization reactors. Major reasons for leaching, other than poor fixation on the support, include solubility of catalyst and/or cocatalyst in the reaction medium and interaction of the cocatalyst (e.g., MAO) with other aluminum alkyls (e.g., TEA, TIBA) that are added to the reactor as scavengers. TIBA is known to modify MAO and increase its solubility in commonly used industrial alkane diluents. Chemical tethering of metallocene on the supports can provide a means to attach the complex covalently to the carrier and, hence, decrease the chance of catalyst leaching. However, such methods of supported metallocene synthesis are not preferred industrially because of the number and complexity of the involved steps, which leads to higher production costs than for supported metallocenes prepared with physical adsorption methods. Furthermore, the highly oxophilic nature of group 4 metallocenes and the fact that the steric and electronic environments of such catalysts are always different from their homogeneous analogs (which can lead to significantly different active sites) are important issues associated with these synthesis methods [41, 65].

Due to the number of variables playing crucial roles, plus the fact that the sequence of reagent addition can also impact the performance of the final supported metallocenes, no universal method has been developed that can provide supported metallocenes with all the required traits. Considerable research needs to be conducted to optimize a particular metallocene.

## 5.3 Types of Cocatalysts

Metallocenes need an activator or cocatalyst for conversion into active olefin polymerization catalysts, regardless of whether they are in homogeneous or heterogeneous form. Major technological developments and fundamental understanding of single-site catalysts have been greatly helped by the discovery of new and more effective cocatalysts. Generally, the cost of a cocatalyst (mainly organometallic compounds of group 13) for group 4 metallocenes is higher than the cost of the catalyst itself, which is another driving force for the development of new, more effective but cheap cocatalysts. Cocatalysts activate metallocenes by extracting one or more of their non-Cp ligands and creating an ion pair in which the transition metal center of the metallocene becomes the cation and the cocatalyst becomes the anion. This process may influence the polymerization process and properties of the obtained polymer.

Since the discovery of metallocenes, different types of cocatalysts have been developed, including aluminum alkyls ($AlR_3$) (e.g., TEA, TIBA, ToA), alumoxanes (e.g., MAO, ethylaluminoxane (EAO), $t$-butylaluminoxane (tBAO), boranes, borates and activating supports. Each of these cocatalysts has a different degree of metallocene activation efficiency. However, MAO can be considered as one of the most commonly studied and used (both in the academia and industry) cocatalysts for metallocenes in ethylene polymerization.

Kaminsky and Sinn [83, 84] showed that MAO is a very effective cocatalyst for olefin polymerization using metallocene catalysts and can sometimes lead to metallocene activities higher than those of traditional ZN catalysts in homogeneous olefin polymerizations [32, 34]. Since then, no other aluminoxane family member (e.g., EAO or tBAO) has been found to be a better cocatalyst for metallocenes than MAO. MAO is produced by the controlled reaction of TMA with water, and is made up of alternate arrangements of aluminum and oxygen atoms, with the free valances being saturated by methyl substituents. The basic structural unit of MAO is $[Al_4O_3Me_6]$ [85, 86]. However, the aluminum atoms are unsaturated in the unit structure, leading to agglomerates of molecules, which can then form cages or clusters of MAO. Although research is still underway on the exact structure of MAO, the consensus of the scientific community seems to be converging on a cage structure with four-coordinated aluminum and three-coordinated oxygen centers, based upon various characterization studies [87]. However, nanotube-like structures [88], linear chains, and cyclic ring structures of MAO are also thought to exist [38].

The molar mass of MAO varies between 700 (corresponding to 12 aluminum atoms) [85, 89] and 18,000 g mol$^{-1}$ (corresponding to aggregates of 150–200 aluminum atoms) [90], and its solubility in aromatic solvents is higher than in aliphatic hydrocarbons [34, 38]. Recently, on the basis of small angle neutron scattering (SANS) and pulsed field gradient spin echo (PFG-SE) NMR measurements, Ghiotto et al. [91] suggested that the molar mass of polymeric MAO is about $1800 \pm 100$ g mol$^{-1}$, corresponding to about 30 aluminum atoms per MAO polymer, and that its hydrodynamic radius is $12.0 \pm 0.3$ Å. When used in a solution process to activate metallocenes, the Al-to-transition metal ratios are of the order of 1000:1 to10,000:1, with some studies also reporting this ratio to be over 300,000:1 [91, 92]. Such high amounts of MAO are needed to shield the active sites from each other and avoid any bimolecular deactivation [27].

The addition of AlR$_3$ compounds (TEA, TIBA, etc.) are reported to increase the solubility of MAO in alkane diluents as well as the activity of metallocenes activated with MAO. This effect has been attributed to the fact that the AlR$_3$ compounds trap free TMA in MAO through Al-alkyl scrambling. TIBA has been found to be better trapping agent for free TMA than TEA due to the fact that mixed alkyl dimers are generated when bulkier AlR$_3$ is added [93].

MAO it is relatively expensive and dangerous, so development of other cocatalysts or activating systems is an active area of research. For a detailed overview of this complex cocatalyst we suggest the recent review of Zijlstra and Harder [93]. The works of Chen and Marks [38] and Boisson et al. [7–9] can be consulted for details about other cocatalyst and activating supports, respectively.

# 6  Impact of Physical Properties of Silica

From the preceding discussion, it should be clear that the physical properties of silica supports are as crucial to the behavior of a supported metallocene catalyst as its surface properties, not least because they impact the distribution of catalyst and cocatalyst throughout the solid particles during catalyst synthesis. In addition, the diffusion of (co)monomer(s), hydrogen, aluminum alkyls, and other components to the active sites during polymerization are crucial to the observed rate of polymerization and to the development of PE molar mass. It has also been shown that even the crystallization of nascent polyolefin chains within the porous support (and thus fragmentation of the particles) can depend on the pore size of the silica. However, despite the importance of these (controllable) properties of silica supports, very few systematic analyses of their importance in the polymerization of ethylene have been carried out. This can pose quite a problem! It is true that a fundamental understanding of the chemistry of active sites is vital to an understanding of how a polymerization takes place. However, if one cannot separate physical phenomena from intrinsic "chemical" phenomena, there is a great risk of confounding the two and thereby clouding our understanding of what is truly happening. Therefore, to exert a maximum degree of control over how a polymerization takes place and what the end result will be, it is important to understand the role of the geometric properties of the support.

## 6.1  Effect of Silica Particle Size

Let us begin by looking at the impact of the most obvious geometric property of the silica support, particle size. Clearly, the size of the support particle is important because (to a great extent) it determines the characteristic length and time scales for heat and mass transfer during polymerization. The characteristic time for diffusion to occur inside a catalyst particle ($\tau$) is equal to the effective diffusivity of the species in question ($D$) divided by the square of the particle radius ($R$), as follows:

$$\tau = \frac{R^2}{D}$$

Thus, if we have a particle with a radius of 10 μm, and one with a radius of 20 μm, it will take four times longer for a molecule of ethylene to diffuse from the surface to the center of the larger particle (all other things being equal). It is, therefore, likely that mass transfer resistance is more important in larger catalyst particles; thus, larger particles can be undersupplied with monomer and polymerize more slowly than smaller ones. This is true not only for monomer, hydrogen, and other reactive species during polymerization, but also for catalyst, alkyl aluminum, and cocatalysts (think of the size of the MAO molecules) during catalyst preparation. For instance, scanning electron microscopy combined with energy dispersive X-ray spectroscopy (SEM-EDX) analysis of catalyst particles of 10–80 μm showed that MAO was uniformly distributed throughout the smaller catalyst particles. For bigger catalyst particles, core–shell distribution was observed, which leads to higher MAO concentrations at the surface than at the particle center [94–96]. This implies that the active sites near the surface of the larger particles probably behave differently from those nearer the center. After 90 min of polymerization (2 bar of propylene and 40°C in toluene), the smaller fragments appeared to be totally fragmented and the rate curves obtained from these particles showed no induction period. On the other hand, the larger particles showed a certain induction time and had unfragmented cores after 90 min of polymerization. Although the reaction conditions considered were not particularly realistic, these experiments suggest that particle size can influence the reaction rate in many ways.

In more recent works, Tisse et al. [97, 98] and Tioni et al. [99] analyzed the impact of the particle size of silica supports on the reaction kinetics and molar mass distribution of ethylene homopolymer and ethylene/1-hexene copolymers. To eliminate the impact of internal particle structure on their observations, the authors sieved a master batch of commercial silica into fractions with particle sizes ranging from 36 to 100 μm. Each sieved fraction was dehydroxylated at 200°C before being used to support $rac$-Et(Ind)$_2$ZrCl$_2$. Elemental analysis of the final catalysts showed very similar metal loadings on all the catalysts, regardless of particle size, suggesting that any observed effects of support size on the observed activity are due to the physical properties of the support, rather than uneven distribution of active sites. EDX analysis of the catalyst particles showed uniform Al distribution throughout the surface and interior of the particles. Slurry phase homopolymerization and ethylene/1-hexene copolymerization with TIBA as scavenger showed that the smaller the particle size of the silica support, the higher the observed rate of polymerization, and the faster the polymerizations reached their maximum activity [97, 98]. It was suggested that these observations were due to enhanced resistance to monomer diffusion on the larger particles, because the particles were identical in every other way. Unfortunately, with this particular metallocene, the MW of the final polymer is largely independent of the monomer pressure so it was not directly possible to use the MWD to reinforce the conclusion. The gas phase polymerizations reported by Tioni et al. [99] showed very similar results in terms of the impact of particle size on the observed rate of polymerization in the gas phase. Once again, it was not possible to use the MWD to prove that mass transfer resistance was the cause of the differences in reaction rates. Tisse et al. [97, 98] also observed that the precontact time between the SMAO-supported metallocenes and aluminum alkyls had a noticeable influence on the observed reaction rate for the 80 μm support, with 1 h contact times giving higher activities than 10 min precontact.

More recently, Bashir et al. [62] carried out similar experiments on three different silicas in both gas and slurry phase polymerizations. In addition to seeing very similar trends in terms of the relationship between higher polymerization rates with smaller particles, they also presented experimental proof of the existence of mass transfer resistance during the early stages of gas phase ethylene/1-hexene copolymerization. The authors analyzed the MWDs of copolymer samples produced at different reaction times and showed that there were visible differences in the width of the MWD curves (i.e., the polydispersity) up to first 30 min of reaction time. Bigger, less active catalyst particles gave PEs with broader MWDs than samples produced with smaller (but more active) catalyst particles during the initial instants of polymerization, as shown in Fig. 9. Given that mass transfer resistance leads to lower monomer concentrations toward the center of the particles, in turn leading to lower observed rates and lower MWs in the particle center, the results proved that mass transfer resistance in silica particles was more significant in larger particles than in small ones. Overlapping of the MWDs after 75 min of reaction time was in agreement with the fact that mass transfer resistance decreases during the reaction because of an increase in external particle surface area of the growing catalyst/polymer particles, as shown by Floyd et al. [100].

In conclusion, a growing body of experimental evidence supports earlier modeling work proposing an impact of particle size on mass transfer rates inside catalyst and polymer particles. However, the vast majority of modeling studies proposed by academic groups have focused on the mass transfer of monomer during the polymerization process. Very little work has centered on the role of mass transfer during the preparation of metallocene catalysts. As discussed (see also Sect. 6.2), the diffusion of large, bulky molecules such as MAO (or other alkyls) is by no means instantaneous and potentially plays an important role in determining the final characteristics of the catalyst, even before the polymerization process begins. Clearly, more work is needed in this area. Let us now turn our attention to the potential impact of the internal structure of the support on catalyst preparation and ethylene polymerization.

## 6.2   Effect of Silica Porosity

Porosity is a general term that collectively refers to silica pore size, pore size distribution, pore volume, and surface area. Just like particle size, silica support porosity also plays a role during the synthesis of supported catalysts and during olefin polymerization. Depending on the method (and conditions) of catalyst synthesis and on catalyst molecular dimensions, porosity can have a strong influence on the distribution of cocatalyst and/or catalyst inside the support particles. It should be noted that pore morphology is typically divided into three families depending on the pore size. Micropores are pores with a diameter of less than 2 nm, mesopores have diameters of 4–200 nm, and macropores have diameters greater than 100 nm [101]. It is claimed that mesoporosity has the strongest influence on the performance of silica-supported catalysts during olefin polymerization [12].

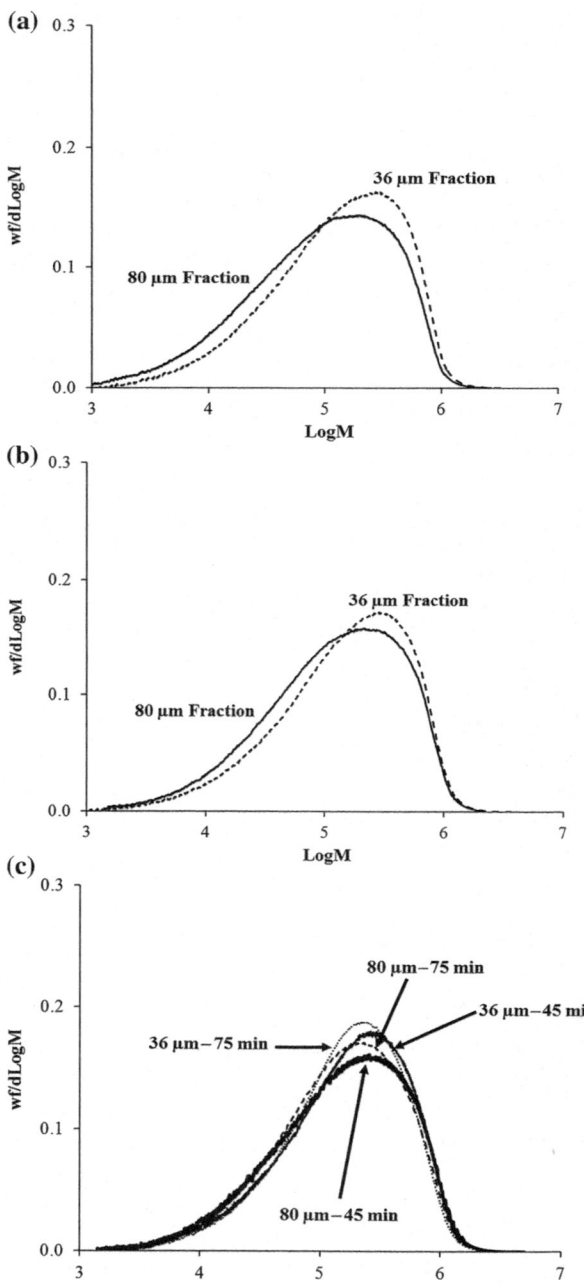

**Fig. 9** MWDs of ethylene/1-hexene copolymer samples produced with silica-supported *rac*-ethylenebis(4,5,6,7-tetrahydro-1-indenyl) zirconium dichloride (THI)/MAO catalyst in a gas phase process. Reactor temperature 80°C, ethylene pressure 8.5 bar. Reaction time: (**a**) 15 min. (**b**) 30 min. (**c**) 45 and 75 min

**Table 1** Physical properties of as-received commercial silica and final supported catalysts [62]

| Silica trade name | Surface area $(A_s)$ [m$^2$/g] | Pore volume $(P_v)$ [mL/g] | Pore diameter $(P_d)$ [Å] | Particle diameter $(d_{50})$ [μm] |
|---|---|---|---|---|
| Grace 948 – as received | 290 | 1.7 | 232 | 60 |
| Grace 948 – plus catalyst | 270 | 0.85 | 111 | |
| PQ MS3040 – as received | 420 | 3.0 | 300 | 45 |
| PQ MS3040 – plus catalyst | 412 | 2.04 | 169 | |
| PQ MS1732 – as received | 536 | 1.4 | 101 | 128 |
| PQ MS1732 – calcined | 507 | 1.32 | 8.8 | |
| PQ MS1732 – plus catalyst | 471 | 0.71 | 56 | |

### 6.2.1 Preparation of Silica-Supported Metallocenes

The size of the support particles can have an influence on the way that different catalyst components are distributed inside the particles. If we refer to the equation for the characteristic diffusion time, it is clear that the diffusivity of different species also plays an important role in determining how species are distributed within catalyst particles. The diffusion coefficient depends on many factors, not least of which are the size of the diffusing molecules and the size of the pores. If we think about the different methods for preparing catalysts (briefly discussed in Sect. 5.2) and the nature of the catalyst components (active site precursors, MAO, alkyl aluminums, etc.), it is not unreasonable to suggest that the nature of the porosity of a given support might be an important consideration! This is clearly a very complex subject, because, for example, the order of addition of the different components could also play a role in altering porosity during the preparation step. Surprisingly, very little attention has been paid to this subject in the open literature, so before discussing the impact of silica porosity on the evolution of polymerization, we briefly discuss a recent study from our group [63] and show that the distribution of metallocene/MAO mixture or MAO alone (depending upon the catalyst synthesis method) can be influenced by the porosity of the silica support.

Bashir et al. [62] used three different commercial silica gels to support ($n$-BuCp)$_2$ZrCl$_2$. The incipient wetness method was used to prepare the catalysts. Each silica was dehydroxylated at 600°C under dynamic vacuum of $10^{-3}$ to $10^{-5}$ mbar before impregnation with the metallocene + MAO mixture. The physical characteristics of the support and the support + catalyst porosity are shown in Table 1. The first thing to notice is that although the specific surface areas of the supports were not influenced (significantly) by dehydroxylation and the act of depositing catalyst in the particles, the pore volumes and pore diameters were

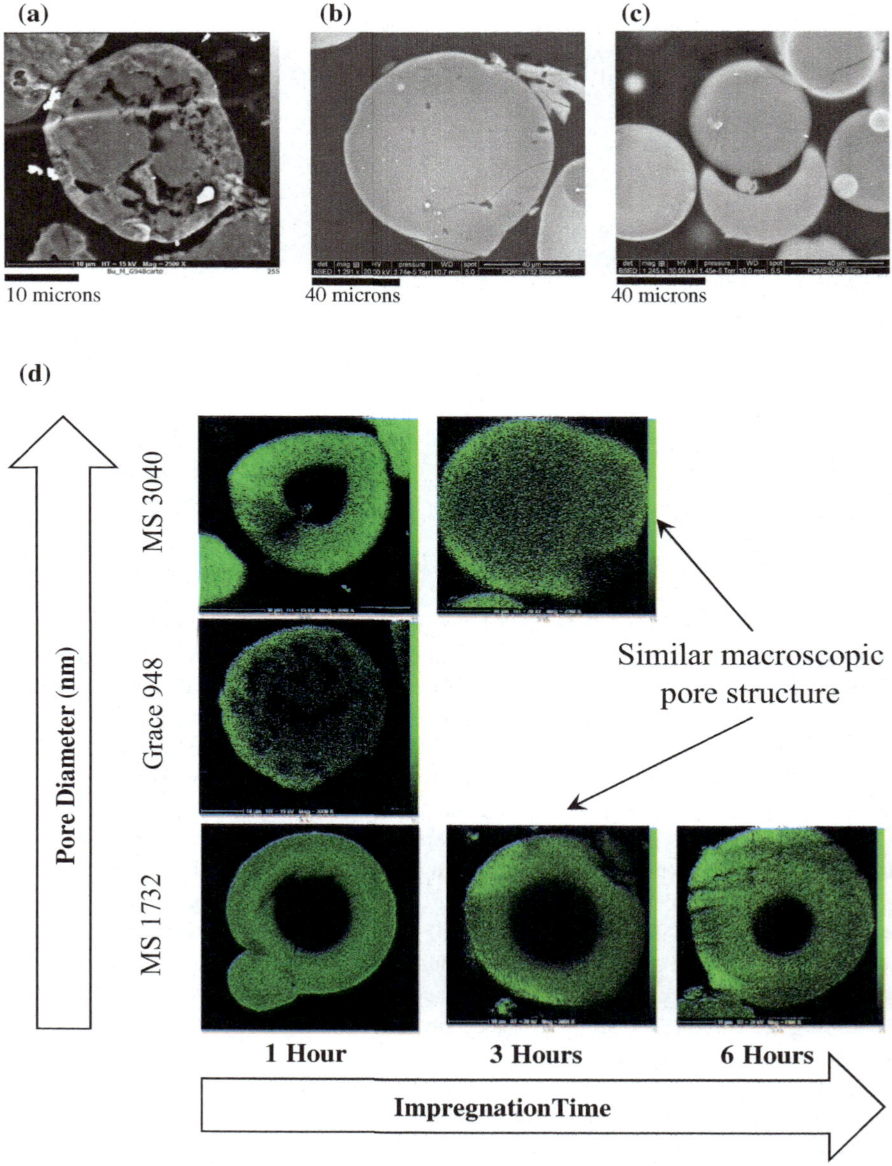

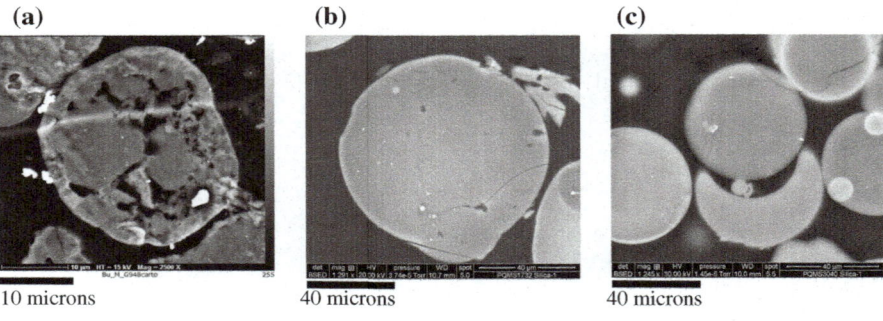

**(d)**

**Fig. 10** (a) Cross-sectional SEM image of (n-BuCp)$_2$ZrCl$_2$/MAO catalyst supported on 63 μm Grace 948 silica. (**b, c**) 63 particles of pure PQ silicas. (**d**) EDX analysis of the same catalysts after different impregnation times. *Green color* indicates aluminum map. Silicon maps are not shown for the sake of clarity

affected. Note that no stirring was used during catalyst synthesis, which allowed the authors to assume that the particle size distribution (PSD) of the final supported catalyst remained similar to that of the original silica.

Full batches of these three silicas were first sieved and the sieved fractions of the same size used to prepare the supported catalysts. In this way, the particle size was controlled while changing the silica porosities. Figure 10a–c shows cross-sectional images of the supported catalysts and Fig. 10d shows EDX mapping of the aluminum (i.e., active sites, because the metallocene was pre-activated with MAO) in the final catalysts with similar particle sizes at different impregnation times. It can be seen that the two PQ silicas have very similar well-connected pore structures but considerably different pore dimensions, whereas Grace 948 silica has interstitial voids indicating that it was prepared by a spray drying process. Figure 10d shows some interesting results. Grace 948, with its relatively macroporous structure, showed an even distribution of Al inside the catalyst particles after 1 h of impregnation. However, for the more homogeneous PQ supports, both catalysts had core–shell aluminum distribution at low impregnation times. Impregnating the larger pore diameter PQMS 3040 silica for up to 3 h gave a fairly even MAO distribution throughout the silica particles. For the low pore diameter PQMS 1732 silica, an impregnation time of up to 6 h was still not enough for uniform distribution of MAO.

These observations can be explained by considering the effect of the presence of interstitial voids, which are generally present in spray-dried silicas such as Grace 948 silica. These interstitial voids start from the particle surface and can penetrate to the particle center, which facilitates the (reactive) diffusion of the metallocene/MAO mixture, and therefore avoids excessive choking of smaller pores with the bulky MAO molecule. On the other hand, PQ silicas do not have such interstitial voids because of the different (emulsion) fabrication process. Although this gives PQMS1732 silica a pore volume similar to that of Grace 948 and very high surface area, the absence of macropores is clearly associated with the core–shell distribution of aluminum.

Smit et al. [59] used full batches of Grace 948 and PQMS 3040 silicas to study the impact of catalyst synthesis method on propylene polymerization rates and found core–shell aluminum distribution in both of their catalysts, which were also prepared with incipient wetness method under similar conditions. However, it is important to highlight that, depending on the Al-to-Zr molar ratio, MAO molecules can aggregate to form clusters with dimensions bigger than the original molecular MAO dimensions; for example, aggregation of two associated ion pairs to an ion quadrupole would be expected to increase the effective radius by a factor of approximately $(2)^{1/3} = 1.26$ [90]. Therefore, even with the same method of catalyst synthesis, it is possible that other aspects of the catalyst formulation can significantly influence MAO distribution inside the support particles, which, in turn, impacts catalyst behavior during polymerization.

## 6.2.2   Kinetics of Polymerization

Various reports have been published on the impact of silica porosity on reaction kinetics. Sano and coworkers [102–104] analyzed the impact of the pore diameter of mesoporous silica supports on the catalytic activities of ethylene homopolymerization with $Cp_2ZrCl_2$ catalyst and propylene homopolymerizations with *rac*-Et(Ind)$_2$ZrCl$_2$ in a slurry process. The basic idea behind the work of these authors was to separate MAO on mesoporous MCM-41 silica of different pore sizes, and use the MAO-impregnated supports in olefin polymerizations to assess the impact of support pore size on catalytic activity, polymer physical and molecular properties, and the number of active species taking part in the polymerization. The word "separate" was used by the authors because it is believed that MAO is a mixture of several oligomers, so using structures of controlled, yet different, pore sizes would allow different oligomers to enter pores of different sizes. For comparison, the authors also used silica gels and silicalite as supports. Before impregnating the supports, they were treated with trimethylchlorosilane to consume the silanol groups present on each silica and, thus, allow the MAO to absorb physically on the silica surface.

Ethylene homopolymerizations were performed for 30 min at 40°C in toluene (no mention of whether or not catalyst leaching was an issue here!). The rate of polymerization was very low when the pore diameter of the MCM-41 support was smallest (i.e., 0.56 nm). Activity increased with increasing pore diameter and showed a maximum value for the catalyst supported on 2.5 nm pore diameter MCM-41 silica. Oddly, the trend was exactly the opposite for silica gel-supported catalysts in both ethylene and propylene (using Et(Ind)$_2$ZrCl$_2$ supported on the same SMAO samples) [103, 104]. However, as no information was provided about the PSD of any of the silicas used, and the chemical natures of the silicas were quite different, it is very difficult to generalize these results. The basic conclusion that one can draw here is that the pore size appears to be important if the other physical *and chemical* properties of the supports are kept (reasonably) constant.

Silveira et al. [105] used two metallocenes, $Cp_2ZrCl_2$ and $(n\text{-}BuCp)_2ZrCl_2$, in the ratio of 1:3 to study the impact of the textural properties of various supports on the supported catalytic activity in ethylene homopolymerization, and on the molecular and physical properties of the final HDPE. The supports included a variety of materials, ranging from conventional Grace 956 silica, pure alumina, various alumino-silicates (e.g., MCM-41, SBA-15, MCM-22), and nonconventional supports such as chrysotile and ITQ-2. Slurry phase ethylene polymerizations were conducted at 60°C for 30 min in a 0.3 L pyrex glass reactor with toluene as diluent and ethylene pressure of 1 bar. It should be noted that, rather than supporting MAO on the catalyst, MAO was fed separately into the reactor in such a way that the Al-to-Zr molar ratio was set to 1000:1 in all reactions. For a given class of support material, catalytic activities were found to be higher for the catalysts with larger pore diameters. The authors attributed this effect to the easy fixation of metallocenes within larger pores, along with easy access of MAO and monomer

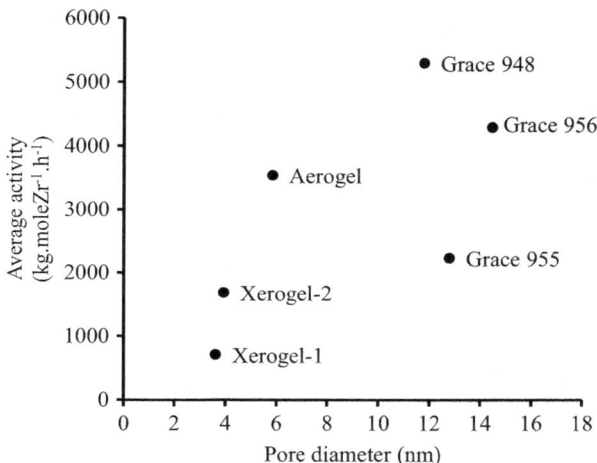

**Fig. 11** Dependence of average catalytic activity on the pore diameter of the final supported catalyst. Data obtained from Silveira et al. (Figure 8) [106]

to the supported metallocene. It is important to mention that no kinetic rate profiles of the reactions were shown, although it would have been helpful for the reader to see differences in catalyst activation as a result of varying support pore diameters. For mesoporous silicas, the authors obtained results similar to those of Sano and coworkers [102–104], but for silica gels they observed the opposite trend. However, it is essential to underline that the approaches used to contact active sites and MAO were very different in the two studies, as was the Al-to-Zr ratio.

It should also be kept in mind that different support materials have different physical characteristics (e.g., resistance to fragmentation) and different surface groups. As discussed, these are extremely important parameters in determining catalyst behavior. For instance, alumina is known to have an amphoteric character due to the presence of both Lewis acidic and basic sites on its surface.

In another work from the same group, Silveira et al. [106] made mixed $Cp_2ZrCl_2$ and $(n\text{-BuCp})_2ZrCl_2$ catalysts (in 1:3 molar ratio) using as supports Grace 948, Grace 955, and Grace 956 silicas, along with two xerogel silicas and one aerogel silica prepared in-house. The polymerization procedure and conditions were kept similar to those used in their previous work [105]. When the average 1 h productivity (in kg PE mol $Zr^{-1}$ $h^{-1}$) was plotted against the pore diameter of each catalyst, a trend could be seen (as shown in Fig. 11). A closer look at Fig. 11 reveals that of the catalysts supported on three Grace silicas, the highest activity was shown by the catalyst supported on Grace 948, which had the lowest pore diameter of the three. The authors attributed the low activities of the catalysts with low pore diameters to the possible formation of inactive bidentate species when the support pore diameter is below 10 nm (i.e., 100 Å). They claimed that in these very small pores, the negative surface curvature keeps the silanol groups very close to each other, which favors the formation of hydrogen bonds between them and, consequently, hinders their removal during heat treatment. However, this seems to imply that the pores are well defined and possibly cylindrical, which was not verified. Thus, one could also add the pore structure/morphology to the list of support properties that impact final catalyst behavior! Note also that when plotted against the particle size of each supported catalyst, the average activity decreased with increasing catalyst particle size.

Wongwaiwattanakul and Jongsomjit [107] analyzed the impact of the pore size of pure silica support on the activity of $rac$-Et(Ind)$_2$ZrCl$_2$ in slurry phase ethylene/ 1-octene copolymerization at 70°C and 3.4 bar ethylene pressure. Dried, modified methylaluminoxane (dMMAO) was supported on each of the materials, and the metallocene complexed with TMA in a toluene solution was added into the reactor separately. The two pure silica supports had unimodal pore size distributions, with one support having an average pore diameter of 13.7 nm and a pore volume of 1.50 mL g$^{-1}$. The second silica support had a pore diameter of 33.8 nm and pore volume of 0.26–1.50 mL g$^{-1}$. The authors found that the silica with the larger pores had an Al content of almost 19 wt% after addition of dMMAO, whereas the silica with smaller pores had only 12 wt% Al. SEM-EDX analysis of the catalyst surface showed an even distribution of Al, but no attempt was made to investigate the internal distribution of Al. Using an Al-to-Zr molar ratio of 1135:1, the authors found that the activity (expressed as kg PE mol Zr$^{-1}$ h$^{-1}$) was higher for the catalyst with the largest pores. The authors proposed that as a result of higher dMMAO loading of the silica support with the largest pore diameter, the concentration of active sites was higher on that catalyst than for the low pore diameter silica support.

Using exactly the same catalyst synthesis procedure and copolymerization conditions, the same group [108] analyzed the effect of the pore size of MCM-41 mesoporous silica supports on the activity of the same metallocene (i.e., $rac$-Et (Ind)$_2$ZrCl$_2$). Once again, dMMAO was used as the cocatalyst. One of the MCM-41 silica supports used in this study possessed unimodal pore size distribution, with small pores having an average diameter of 2 nm. The other MCM-41 supports had a bimodal pore size distribution, with large and small pores with average pore diameters of 5 and 6 nm. The unimodal support had a specific surface area of 864 m$^2$ g$^{-1}$, which is twice the specific surface area of the two other bimodal supports. The authors found that the catalysts supported on the silicas with bimodal pore size distribution (those having both small and large pores but smaller surface area – about half that of the unimodally distributed support) were approximately 30% more active than the same catalyst supported on the silica with unimodal pore size distribution. The authors proposed that, although the smaller pore size led to higher surface area and better dispersion of cocatalyst and catalyst, mass transfer resistance to monomer(s) transport within the pores at the reaction startup and during polymerization was higher, which reduced the activity of the final supported catalyst. They further claimed that the support with bimodal pore size distribution provided the benefit of good active site distribution (due to small pores) and reduced mass transfer resistance to monomer(s) transport (due to large pores), which led to higher catalytic activities. However, the authors also used thermogravimetric analysis to show that the interaction of dMMAO with the unimodal small pore sized MCM-41 silica support was significantly stronger than its interaction with bimodal large pore sized MCM-41 supports. Given that the strongly bound cocatalyst is less effective, it is more likely that this is the reason for the difference in activities, rather than monomer mass transfer (no clear justification was given for assuming monomer mass transfer limitations were important). Furthermore, the polydispersity of the MWD of the polymers produced with larger and bimodal pores was about twice that of the copolymer produced with catalyst supported on small pore sized unimodal silica support. This result also indicates that differences in the nature of the active sites of the different catalysts were the origin of the observed differences in catalytic activities.

Tisse et al. [98] evaluated the impact of silica support porosity and PSD of silica supports on the activity of supported $rac$-Et(Ind)$_2$ZrCl$_2$ metallocene in slurry homo- and copolymerizations. This work used two types of activation processes for $rac$-Et (Ind)$_2$ZrCl$_2$: (1) activating supports with attached aluminum and fluoride species, giving enough acidity to the supports so that they can activate the metallocene for olefin polymerization [9]; and (2) silicas impregnated with MAO (SMAO) for comparison and to correlate the observed trends with the type and size of activator used in addition to the support properties. All the supported catalysts were prepared on commercial silicas with pore volumes of 1–3.2 mL g$^{-1}$, pore diameters of 3.7–40.0 nm, and surface areas of 290–800 m$^2$ g$^{-1}$.

Although the authors identified no clear trend between the rate of polymerization and support properties such as pore volume, pore diameter, and surface area, a new look at this information shows that the results are consistent with those of Sano et al. [102].

Figure 12 shows the tabular data of Tisse et al. [98] in graphical form, correlating the pore volume and pore diameter of each support with the reported average activity and surface activity (obtained by dividing the reported average activity with the corresponding reported surface area of the support, because the surface area of final catalysts was not given by the authors). Figure 12a, b demonstrates that both the average activity and surface activity show a maximum value at a pore volume of about 2 mL g$^{-1}$, after which they decrease to an almost constant value. When plotted against pore diameter, a similar trend can be noticed with initial rise in activities (maximum at about 25 nm) followed by decreasing activities with increasing pore diameter up to 40 nm (see Fig. 12c, d). These graphs indicate that the pore volume and pore diameter of the silica-supported catalyst are very important physical parameters and should be optimized in such a way that the catalyst is not completely inactive (e.g., see the first point in Fig. 12a, d) or functioning at the lowest activity within the selected range. To supplement these observations, Fig. 12e shows the plot of activity versus pore diameter from the publication by Sano et al. [102]. A similar dependence of catalytic activity can be seen for a similar range of pore sizes for the same metallocene, but supported on different silica, in slurry phase ethylene polymerization under different conditions. The most probable explanation for this type of dependence of catalytic activity on the pore volume and pore diameter of the silica supports is that, once inside the reactor, fragmentation of the supported catalyst with low pore volume occurs faster than that of catalyst with higher pore volume (assuming similar metal loadings), which leads to higher activities of low pore volume (and diameter) catalysts. However, we should note

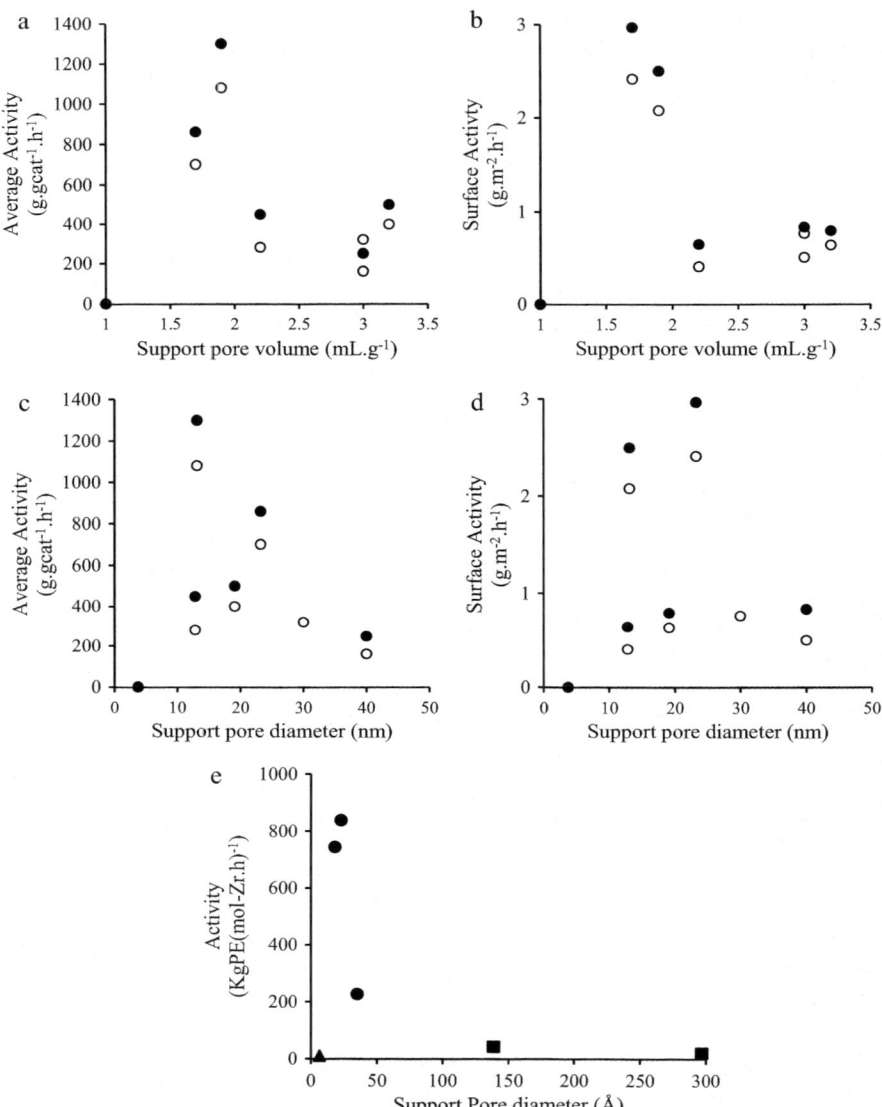

**Fig. 12** (**a–d**) Graphical presentation of the experimental data reported by Tisse et al. [98] on average activities, pore volumes, and pore diameters of the activating support-based catalysts. (**e**) The data of Sano et al. [102], presented with permission. *Empty circles* show homopolymerizations and *filled circles* show copolymerizations; different markers indicate the different materials used

that no comparison of the kinetic profiles of such catalysts was provided by the authors. In addition, Tisse et al. [98] used silicas from various manufacturers, made with different production processes. Given the range of materials used, it is likely that they had different pore structures and fragility levels. Nevertheless, the similarity between the two sets of data gives pause for thought!

Kumkaew et al. [109, 110], carried out similar studies, but in gas phase polymerization. They focused on an analysis of the effect of pore diameter of mesoporous molecular sieves and silicates on the reaction rate and comonomer incorporation. Molecular sieves with a broad range of pore diameters were mostly used as supports, but a silicate support was also looked at. First, MAO was supported, followed by grafting of $(n\text{-BuCp})_2\text{ZrCl}_2$ at room temperature for 4 h. Then, the catalysts were vacuum dried at room temperature for several hours to provide the free flowing supported catalysts. For the catalysts supported on molecular sieves, the authors found that the smaller the pore diameter, the higher were the instantaneous and average activities in comparison with the catalysts having larger pore diameters, for both homo- and copolymerization. This difference in the kinetic profiles was more pronounced at higher reaction temperatures than at lower temperatures. It is important to mention here that the particle sizes of the used supports were not kept constant in this study, so it is difficult to judge whether pore size alone can account for these differences. Furthermore, the authors also noticed a difference in comonomer incorporation for the different supports. Temperature rising elution fractionation (TREF) results for copolymer samples obtained using catalysts of 2.6–20 nm pore diameter showed at least two distinct peaks, one at 55–70°C and a second at about 98°C. With increasing catalyst pore diameter, the low temperature peak (i.e., in the range 55–70°C) became dominant, indicating enhanced 1-hexene incorporation and reduced formation of homopolymer (note that the activity decreased with increased catalyst pore diameter). The presence of at least two distinct peaks in the TREF analysis strongly suggests that there were at least two types of active site. The authors suggested that the nature of active sites formed on a supported catalyst can be affected by its pore size. Because the catalyst with smaller pore diameters showed higher amounts of homopolymer than catalysts with bigger pores, the authors suggested that the confined (i.e., small) pore size probably affects the structure of supported MAO and, consequently, affects the nature of interactions between MAO and $(n\text{-BuCp})_2\text{ZrCl}_2$, which leads to active species that produce higher amounts of homopolymer. As the support pore size increases, the effect of confinement reduces and leads to a different type of active site that can incorporate greater amounts of 1-hexene.

The catalyst particle size also plays a crucial role in determining catalyst behavior during ethylene polymerization. This aspect is ignored in many reported studies and it is difficult to know whether or not the catalyst particle sizes were kept constant while studying the impact of catalyst porosity on reaction kinetics. Recently, our group [62] studied silica-supported $(n\text{-BuCp})_2\text{ZrCl}_2/\text{MAO}$ catalysts of different porosities but similar particle sizes, prepared as described in Sect. 6.2.1. It can be seen from Fig. 13a, b that the sieving operation provided silica fractions of very similar PSDs but different porosities (for representative porosities see

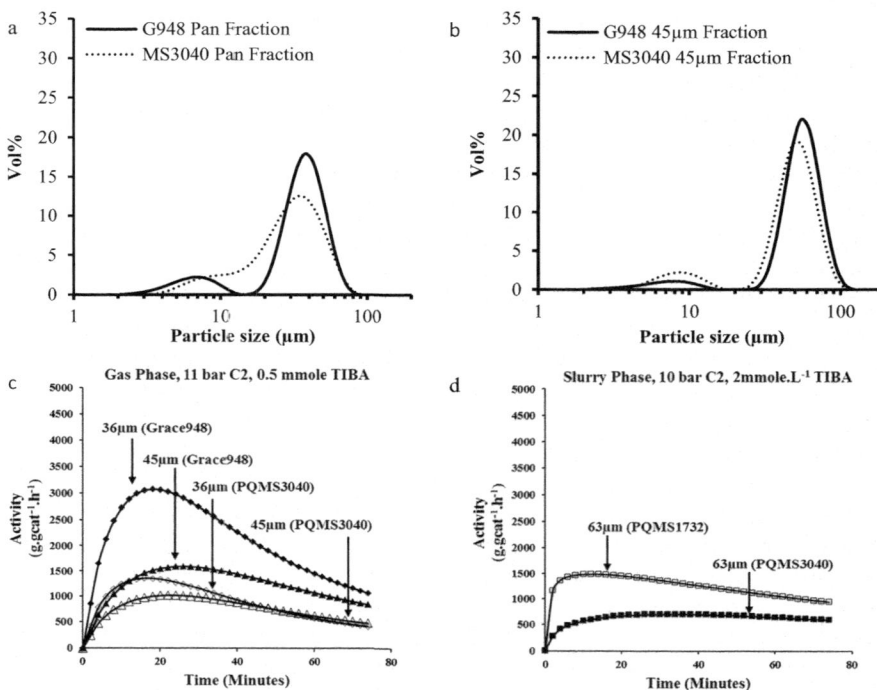

**Fig. 13** (**a**, **b**) Particle size distribution of sieved fractions of Grace 948 and PQMS 3040 silica. (**c**) Gas phase copolymerization kinetics of ($n$-BuCp)$_2$ZrCl$_2$/MAO catalyst supported on 36 and 45 μm sieved fractions of Grace 948 and PQMS3040 silica, (**d**) Slurry phase copolymerization kinetics of ($n$-BuCp)$_2$ZrCl$_2$/MAO catalyst supported on 63 μm sieved fractions of PQMS1732 and PQMS3040 silica. Initial volume of 1-hexene added into the reactor was 3 mL

Table 1). No stirring was applied during catalyst synthesis, so it can be reasonably assumed that the PSDs of the final supported catalysts were similar to the used silica fractions. Figure 13c, d shows that the smaller the pore diameter (or pore volume) of the silica-supported metallocene/MAO catalyst, the higher the instantaneous polymerization rate in both types of polymerizations. Furthermore, low activation and deactivation rates of the high pore diameter catalysts (i.e., those based on PQMS3040 silica) are also evident. These results are in full agreement with the work of Kumkaew et al. [109, 110], and Paredes et al. [111], who used the same metallocene/MAO catalyst but supported on molecular sieves and SBA-15 silica-based mesostructured materials, respectively. The most probable reason for the higher activities of low pore diameter catalysts (i.e., PQMS1732 and Grace 948 silica-based) is faster pore filling during the initial instants of polymerization (as compared with the pore filling of PQMS3040 silica-based catalysts), leading to faster particle fragmentation of the catalysts. It is important to mention here that similar trends were also observed by our group in gas and slurry phase ethylene homopolymerizations using the same supported catalysts [63].

In a slightly different vein, Tioni et al. [112] conducted very rapid reactions (0.3–180 s) in a specially designed stop-flow reactor. Grace 948 silica-supported ($n$-BuCp)$_2$ZrCl$_2$ and $rac$-Et(Ind)$_2$ZrCl$_2$ catalysts were prepared by first impregnating the silica with MAO, followed by metallocene grafting. Gas phase ethylene homopolymerizations and ethylene/1-butene copolymerizations were conducted at 9 bar ethylene pressure and 80°C. The results showed that short reactions of a fraction of a second produced polymer with a melting point of about 118°C. As the reaction continued to 180 s, the crystalline polymer fraction became significant, leading to a melting temperature of 131°C. PE crystallization peaks also showed a similar dependence on polymerization time. The authors attributed this observation to the pore confinement effect, which perturbs the crystallization of nascent polymer chains. As the reaction continues, the confinement effect vanishes because of fragmentation of the support. Tioni et al. based their reasoning on a set of experiments by the group of Woo [113, 114], who deposited PE in the pores of alumina pellets with very well-defined sizes, and showed that there was a significant depression of the melting and crystallization temperatures for pores smaller than 15–20 nm. This result indicates that confinement of nascent polymer in small pores leads to different physical properties, and thus undoubtedly has an effect on the fragmentation and effective activation of supported catalysts.

# 7   Conclusion

This chapter has focused on the impact of the geometric properties of silica on the behavior of supported metallocene catalysts. We have simplified certain explanations and concentrated more on the impact of the supports on kinetics than on material properties in an effort to avoid clouding the issues. We have also talked about issues related to mass transfer, and left heat transfer considerations to the side for the simple reaction that heat transfer has less to do with the internal structure of particles than with the rate of reaction and overall particle size.

We have attempted to show that the act of supporting a metallocene involves more than simply choosing an appropriate formulation. There are many different ways to "put" a metallocene precursor onto a solid support, different activators with different sizes and structures, and many different types of supports. Although we concentrated only on undoped silica gel supports, it should be clear that how the silica is pretreated has a different impact on different metallocenes. In addition, large bulky molecules such as MAO might not diffuse uniformly through a support particle for several reasons, depending on the time taken for this step, temperature, and structure of the catalyst support.

Perhaps most frustratingly, a survey of the rather limited number of studies that examined the importance of pore size and pore size distribution showed that these quantities also have an impact on the polymerization. However, with one or two exceptions, authors tended to vary pore size and/or surface areas in their studies without bothering to control the particle size. In addition, they compared supports with different geometric parameters, but also different compositions. All of this means that it is difficult to draw precise conclusions about the role that particle porosity plays in terms of reaction rates, catalyst activation, particle morphology, and so on. Furthermore, reaction rate profiles are rarely provided by authors. Such profiles would clearly help in determining the role of support properties in catalyst activation (one of the most important stages in olefin polymerization via heterogeneous catalysts).

It appears that there is a range of pore diameters that lead to high activity. However, proof remains elusive. This is bad news for polymer producers, but good news for researchers who find themselves faced with a real challenge in terms of furthering our understanding in this area.

# References

1. Soares JBP, McKenna TFL (2012) Introduction to polyolefins. Polyolefin reaction engineering. Wiley-VCH, Weinheim
2. Peacock AJ (2000) Handbook of polyethylene structures, properties and applications. Marcel Dekker, New York
3. Karian HG (ed) (2003) Handbook of polypropylene and polypropylene composites2nd edn. Marcel Dekker, New York
4. Odian G (2004) Stereochemistry of polymerization. Principles of polymerization. Wiley, Hoboken
5. Soares JBP, McKenna TFL (2012) Polyolefin reactors and processes. Polyolefin reaction engineering. Wiley-VCH, Weinheim
6. Galli P, Vecellio G (2001) Technology: driving force behind innovation and growth of polyolefins. Prog Polym Sci 26:1287
7. Prades F, Boisson C, Spitz R, Razavi A (2005) Activating supports for metallocene catalysis. Patent US7759271B2
8. Prades F, Spitz R, Boisson C, Sirol S, Razavi A (2007) Transition metal complexes supported on activating fluorinated support. Patent WO2007014889 A1
9. Prades F, Broyer JP, Belaid I, Boyron O, Miserque O, Spitz R, Boisson C (2013) Borate and MAO free activating supports for metallocene complexes. ACS Catal 3:2288
10. McInnis JP, Delferro M, Marks TJ (2014) Multinuclear group 4 catalysis: olefin polymerization pathways modified by strong metal-metal cooperative effects. Acc Chem Res 47:2545
11. Nikolaeva MI, Mikenas TB, Matsko MA, Echevskaya LG, Zakharov VA (2010) Heterogeneity of active sites of Ziegler-Natta catalysts: the effect of catalyst composition on the MWD of polyethylene. J Appl Polym Sci 115:2432
12. Mcdaniel MP (2011) Influence of catalyst porosity on ethylene polymerization. ACS Catal 1:1394
13. Soares JBP, McKenna TFL (2012) Particle growth and single particle modeling. Polyolefin reaction engineering. Wiley-VCH, Weinheim
14. Zheng X, Smit M, Chadwick JC, Loos J (2005) Fragmentation behavior of silica-supported metallocene/MAO catalyst in the early stages of olefin polymerization. Macromolecules 38:4673
15. Zheng X, Pimplapure MS, Weickert G, Loos J (2006) Influence of copolymerization on fragmentation behavior using Ziegler-Natta catalysts. Macromol Rapid Commun 27:15

16. Pater JTM, Weickert G, Loos J, van Swaaij WPM (2001) High precision prepolymerization of propylene at extremely low reaction rates-kinetics and morphology. Chem Eng Sci 56:4107

17. Noristi L, Marchetti E, Baruzzi G, Sgarzi P (1994) Investigation on the particle growth mechanism in propylene polymerization with MgCl2-supported ziegler-Natta catalysts. J Polym Sci A Polym Chem 32:3047

18. McKenna TFL, Di Martino A, Weickert G, Soares JBP (2010) Particle growth during the polymerisation of olefins on supported catalysts, 1-nascent polymer structures. Macromol React Eng 4:40

19. McKenna TFL, Tioni E, Ranieri MM, Alizadeh A, Boisson C, Monteil V (2013) Catalytic olefin polymerisation at short times: studies using specially adapted reactors. Can J Chem Eng 91:669

20. Ferrero MA, Koffi E, Sommer R, Conner WC (1992) Characterization of the changes in the initial morphology for MgCl2-supported Ziegler-Natta polymerization catalysts. J Polym Sci A Polym Chem 30:2131

21. Ferrero MA, Sommer R, Spanne P, Jones KW, Conner WC (1993) X-ray microtomography studies of nascent polyolefin particles polymerized over magnesium chloride-supported catalysts. J Polym Sci A Polym Chem 31:2507

22. Knoke S, Ferrari D, Tesche B, Fink G (2003) Microkinetic videomicroscopic analysis of olefin polymerization with a supported metallocene catalyst. Angew Chem Int Ed 42:5090

23. Jang YJ, Naundorf C, Klapper M, Mullen K (2005) Study of the fragmentation process of different supports for metallocenes by laser scanning confocal fluorescence microscopy (LSCFM). Macromol Chem Phys 206:2027

24. Grof Z, Kosek J, Marek M (2005) Principles of the morphogenesis of polyolefin particles. Ind Eng Chem Res 44:2389

25. Grof Z, Kosek J, Marek M (2005) Modeling of morphogenesis of growing polyolefin particles. AICHE J 51:2048

26. Gambarotta S (2003) Vanadium-based Ziegler-Natta: challenges, promises, problems. Coord Chem Rev 237:229

27. Soares JBP, McKenna TFL (2012) Polymerization catalysis and mechanism. Polyolefin reaction engineering. Wiley-VCH, Weinheim

28. Billmeyer FW (1984) Ionic and coordination chain (addition) polymerization. Textbook of polymer science3rd edn. Wiley Interscience, New York

29. Chadwick JC, Garoff T, Severn JR (2008) Traditional heterogeneous catalysts. Tailor-made polymers. Wiley-VCH, Weinheim

30. McDaniel MP (2010) A review of the Phillips supported chromium catalyst and its commercial use for ethylene polymerization. In: Gates BC, Knozinger H, Jentoft FC (eds) Advances in catalysis, vol 53. Academic, San Diego

31. McDaniel MP (2008) Review of the Phillips chromium catalyst for ethylene polymerization. Handbook of heterogeneous catalysis. Wiley-VCH, Weinheim

32. Kaminsky W (1998) Highly active metallocene catalysts for olefin polymerization. J Chem Soc Dalton Trans 1998:1413

33. Kaminsky W, Kopf J, Sinn H, Vollmer HJ (1976) Extrem verzerrte Bindungswinkel bei Organozirconium-Verbindungen, die gegen Ethylen aktiv sind. Angew Chem 88:688

34. Kaminsky W (2012) Discovery of methylaluminoxane as cocatalyst for olefin polymerization. Macromolecules 45:3289

35. Sinn H, Kaminsky W (1980) Ziegler-Natta catalysis. In: Stone FGA, West R (eds) Advances in organometallic chemistry, vol 18. Academic, New York

36. Breslow DS, Newburg NR (1957) Bis-(cyclopentadienyl)-titanium dichloride-alkylaluminum complexes as catalysts for the polymerization of ethylene. J Am Chem Soc 79:5072

37. Natta G, Pino P, Mazzanti G, Giannini U (1957) A crystallizable organometallic complex containing titanium and aluminum. J Am Chem Soc 79:2975

38. Chen EY-X, Marks TJ (2000) Cocatalysts for metal-Catalyzed olefin polymerization: activators, activation processes, and structure-activity relationships. Chem Rev 100:1391
39. Kaminsky W (2016) Production of polyolefins by metallocene catalysts and their recycling by pyrolysis. Macromol Symp 360:10
40. Buffet JC, Wanna N, Arnold TAQ, Gibson EK, Wells PP, Wang Q, Tantirungrotechai J, O'Hare D (2015) Highly tunable catalyst supports for single-site ethylene polymerization. Chem Mater 27:1495
41. Severn JR, Chadwick JC, Duchateau R, Friederichs N (2005) "Bound but not gagged" immobilizing single-site α-olefin polymerization catalysts. Chem Rev 105:4073
42. Soares JBP, McKenna TFL (2012) Polymerization kinetics. Polyolefin reaction engineering. Wiley-VCH, Weinheim
43. Stürzel M, Mihan S, Mülhaupt R (2016) From multisite polymerization catalysis to sustainable materials and all-polyolefin composites. Chem Rev 116:1398
44. Campos JM, Lourenco JP, Cramail H, Ribeiro MR (2012) Nanostructured silica materials in olefin polymerisation: from catalytic behaviour to polymer characteristics. Prog Polym Sci 37:1764
45. Heurtefeu B, Bouilhac C, Cloute É, Taton D, Deffieux A, Cramail H (2011) Polymer support of "single-site" catalysts for heterogeneous olefin polymerization. Prog Polym Sci 36:89
46. Hlatky GG (2000) Heterogeneous single-site catalysts for olefin polymerization. Chem Rev 100:1347
47. Klapper M, Joe D, Nietzel S, Krumpfer JW, Müllen K (2014) Olefin polymerization with supported catalysts as an exercise in nanotechnology. Chem Mater 26:802
48. Ribeiro MR, Deffieux A, Portela MF (1997) Supported metallocene complexes for ethylene and propylene polymerizations: preparation and activity. Ind Eng Chem Res 36:1224
49. Martin JL, Thorn MG, Mcdaniel MP, Jensen MD, Yang Q, Deslauriers PJ, Kertok ME (2007) Polymerization catalysts and process for producing bimodal polymers in a single reactor. Patent US7312283B2
50. Collins RA, Russell AF, Mountford P (2015) Group 4 metal complexes for homogeneous olefin polymerisation: a short tutorial review. Appl Petrochem Res 5:153
51. McKnight AL, Waymouth RM (1998) Group 4 ansa-cyclopentadienyl-amido catalysts for olefin polymerization. Chem Rev 98:2587
52. Theopold KH, Heintz RA, Noh SK, Thomas BJ (1992) Homogeneous chromium catalysts for olefin polymerization. Homogeneous transition metal catalyzed reactions, vol 230. American Chemical Society, Washington
53. Theopold KH (1998) Homogeneous chromium catalysts for olefin polymerization. Eur J Inorg Chem 1998:15
54. Severn JR (2008) Methylaluminoxane (MAO), silica and a complex: the "holy trinity" of supported single-site catalyst. Tailor-made polymers. Wiley-VCH, Weinheim
55. Severn JR, Chadwick JC (2013) Immobilisation of homogeneous olefin polymerisation catalysts. Factors influencing activity and stability. Dalton Trans 42:8979
56. Zhuravlev LT (2000) The surface chemistry of amorphous silica. Zhuravlev model. Colloids Surf A Physicochem Eng Asp 173:1
57. Atiqullah M, Akhtar MN, Moman AA, Abu-Raqabah AH, Palackal SJ, Al-Muallem HA, Hamed OM (2007) Influence of silica calcination temperature on the performance of supported catalyst SiO2-nBuSnCl3/MAO/(nBuCp)2ZrCl2 polymerizing ethylene without separately feeding the MAO cocatalyst. Appl Catal A Gen 320:134
58. dos Santos JHZ, Krug C, da Rosa MB, Stedile FC, Dupont J, de Camargo Forte M (1999) The effect of silica dehydroxylation temperature on the activity of SiO2-supported zirconocene catalysts. J Mol Catal A Chem 139:199
59. Smit M, Zheng X, Loos J, Chadwick JC, Koning CE (2005) Effects of methylaluminoxane immobilization on silica on the performance of zirconocene catalysts in propylene polymerization. J Polym Sci A Polym Chem 43:2734

60. Van Grieken R, Carrero A, Suarez I, Paredes B (2007) Ethylene polymerization over supported MAO/(nBuCp)2ZrCl2 catalysts: influence of support properties. Eur Polym J 43:1267
61. Hammawa H, Wanke SE (2007) Influence of support friability and concentration of α-olefins on gas-phase ethylene polymerization over polymer-supported metallocene/methylaluminoxane catalysts. J Appl Polym Sci 104:514
62. Bashir MA (2016) Impact of physical properties of silica supported metallocenes on their ethylene polymerisation kinetics and polyethylene properties. Université Claude Bernard Lyon-1, Villeurbanne
63. Bashir MA, Vancompernolle T, Gauvin RM, Delevoye L, Merle N, Monteil V, Taoufik M, McKenna TFL, Boisson C (2016) Silica/MAO/(n-BuCp)2ZrCl2 catalyst: effect of support dehydroxylation temperature on the grafting of MAO and ethylene polymerization. Cat Sci Technol 6:2962
64. Alt G (1999) The heterogenization of homogeneous metallocene catalysts for olefin polymerization. J Chem Soc, Dalton Trans 1999:1703
65. Choi Y, Soares JB (2012) Supported single-site catalysts for slurry and gas-phase olefin polymerisation. Can J Chem Eng 90:646
66. dos Santos JHZ, Larentis A, da Rosa MB, Krug C, Baumvol IJR, Dupont J, Stedile FC, de Camargo Forte M (1999) Optimization of a silica supported bis(butylcyclopentadienyl)-zirconium dichloride catalyst for ethylene polymerization. Macromol Chem Phys 200:751
67. dos Santos JHZ, Dorneles S, Stedile FC, Dupont J, de Camargo Forte MM, Baumvol IJR (1997) Silica supported zirconocenes and Al-based cocatalysts: surface metal loading and catalytic activity. Macromol Chem Phys 198:3529
68. Chang M (1991) Olefin polymerization catalyst from trialkylaluminum mixture, silica gel and a metallocene. Patent US5006500A
69. Chang M (1992) Method for preparing a silica gel supported metallocene-alumoxane catalyst. Patent US5086025A
70. Chang M (1993) Supported catalyst for 1-olefin(s) (co)polymerization. Patent US5238892A
71. Simplicio LMT, Costa FG, Boaventura JS, Sales EA, Brandao ST (2004) Study of some parameters on the zirconocene immobilization over silica. J Mol Catal A Chem 216:45
72. Lee DH, Shin SY, Lee DH (1995) Ethylene polymerization with metallocene and trimethylaluminumtreated silica. Macromol Symp 97:195
73. Takahashi T (1991) Process for producing ethylene copolymers. Patent US5026797A
74. Welborn HC (1989) Supported polymerization catalyst. Patent US4808561A
75. Akhtar MN, Atiqullah M, Moman AA, Abu-Raqabah AH, Ahmed N (2008) Supported (nBuCp)2ZrCl2 catalysts: effects of selected Lewis acid organotin silica surface modifiers on ethylene polymerization. Macromol React Eng 2:339
76. Atiqullah M, Anantawaraskul S, Emwas AH, Al-Harthi MA, Hussain I, Ul-Hamid A, Hossaen A (2014) Silica-supported (nBuCp)2ZrCl2: effect of catalyst active center distribution on ethylene-1-hexene copolymerization. Polym Int 63:955
77. Chao C, Pratchayawutthirat W, Praserthdam P, Shiono T, Rempel GL (2002) Copolymerization of ethylene and propylene using silicon tetrachloride-modified silica/MAO with et [Ind]2ZrCl2 metallocene catalyst. Macromol Rapid Commun 23:672
78. Jongsomjit B, Kaewkrajang P, Wanke SE, Praserthdam PA (2004) Comparative study of ethylene/α-olefin copolymerization with silane-modified silica-supported MAO using zirconocene catalysts. Catal Lett 94:205
79. Soga K, Shiono T, Kim HJ (1993) Activation of SiO2-supported zirconocene catalysts by common trialkylaluminiums. Makromol Chem 194:3499
80. Kamfjord T, Wester TS, Rytter E (1998) Supported metallocene catalysts prepared by impregnation of MAO modified silica by a metallocene/monomer solution. Macromol Rapid Commun 19:505

81. Moroz BL, Semikolenova NV, Nosov AV, Zakharov VA, Nagy S, O'Reilly NJ (1998) Silica-supported zirconocene catalysts: preparation, characterization and activity in ethylene polymerization. J Mol Catal A Chem 130:121
82. Rytter E, Ott M (2001) Supported metallocene catalysts prepared by impregnation of silica with metallocene/aluminoxane/1-hexene solutions. Macromol Rapid Commun 22:1427
83. Sinn H, Kaminsky W, Vollmer HJ, Woldt R (1980) "Lebende Polymere" bei Ziegler-Katalysatoren extremer Produktivitat. Angew Chem 92:396
84. Kaminsky W, Miri M, Sinn H, Woldt R (1983) Bis(cyclopentadienyl)zirkon-verbindungen und aluminoxan als Ziegler-Katalysatoren für die polymerisation und copolymerisation von olefinen. Makromol Chem, Rapid Commun 4:417
85. Sinn H (1995) Proposals for structure and effect of methylalumoxane based on mass balances and phase separation experiments. Macromol Symp 97:27
86. Koide Y, Bott SG, Barron AR (1996) Alumoxanes as cocatalysts in the palladium-catalyzed copolymerization of carbon monoxide and ethylene: genesis of a structure-activity relationship. Organometallics 15:2213
87. Zijlstra HS, Stuart MCA, Harder S (2015) Structural investigation of methylalumoxane using transmission electron microscopy. Macromolecules 48:5116
88. Linnolahti M, Severn J, Pakkanen T (2008) Formation of nanotubular methylaluminoxanes and the nature of the active species in single-site α-olefin polymerization catalysis. Angew Chem Int Ed 47:9279
89. Ystenes M, Eilertsen JL, Liu J, Ott M, Rytter E, Stovneng JA (2000) Experimental and theoretical investigations of the structure of methylaluminoxane (MAO) cocatalysts for olefin polymerization. J Polym Sci A Polym Chem 38:3106
90. Babushkin DE, Brintzinger HH (2002) Activation of dimethyl zirconocene by methylaluminoxane (MAO) size estimate for Me-MAO-anions by pulsed field-gradient NMR. J Am Chem Soc 124:12869
91. Ghiotto F, Pateraki C, Tanskanen J, Severn JR, Luehmann N, Kusmin A, Stellbrink J, Linnolahti M, Bochmann M (2013) Probing the structure of methylalumoxane (MAO) by a combined chemical, spectroscopic, neutron scattering, and computational approach. Organometallics 32:3354
92. Matsui S, Mitani M, Saito J, Tohi Y, Makio H, Matsukawa N, Takagi Y, Tsuru K, Nitabaru M, Nakano T, Tanaka H, Kashiwa N, Fujita T (2001) A family of zirconium complexes having two phenoxy-imine chelate ligands for olefin polymerization. J Am Chem Soc 123:6847
93. Zijlstra HS, Harder S (2015) Methylalumoxane-history, production, properties, and applications. Eur J Inorg Chem 2015:19
94. Fink G, Steinmetz B, Zechlin J, Przybyla C, Tesche B (2000) Propene polymerization with silica-supported metallocene/MAO catalysts. Chem Rev 100:1377
95. Goretzki R, Fink G, Tesche B, Steinmetz B, Rieger R, Uzick W (1999) Unusual ethylene polymerization results with metallocene catalysts supported on silica. J Polym Sci A Polym Chem 37:677
96. Steinmetz B, Tesche B, Przybyla C, Zechlin J, Fink G (1997) Polypropylene growth on silica-supported metallocene catalysts: a microscopic study to explain kinetic behavior especially in early polymerization stages. Acta Polym 48:392
97. Tisse VF, Prades F, Briquel R, Boisson C, McKenna TFL (2010) Role of silica properties in the polymerisation of ethylene using supported metallocene catalysts. Macromol Chem Phys 211:91
98. Tisse VF, Briquel RM, McKenna TFL (2009) Influence of silica support size on the polymerisation of ethylene using a supported metallocene catalyst. Macromol Symp 285:45
99. Tioni E, Broyer JP, Monteil V, McKenna T (2012) Influence of reaction conditions on catalyst behavior during the early stages of gas phase ethylene homo- and copolymerization. Ind Eng Chem Res 51:14673

100. Floyd S, Heiskanen T, Taylor TW, Mann GE, Ray WH (1987) Polymerization of olefins through heterogeneous catalysis. VI. Effect of particle heat and mass transfer on polymerization behavior and polymer properties. J Appl Polym Sci 33:1021
101. Webb SW, Weist EL, Chiovetta MG, Laurence RL, Conner WC (1991) Morphological influences in the gas phase polymerization of ethylene by silica supported chromium oxide catalysts. Can J Chem Eng 69:665
102. Sano T, Doi K, Hagimoto H, Wang Z, Uozumi T, Soga K (1999) Adsorptive separation of methylalumoxane by mesoporous molecular sieve MCM-41. Chem Commun 1999:733
103. Sano T, Hagimoto H, Sumiya S, Naito Y, Oumi Y, Uozumi T, Soga K (2001) Application of porous inorganic materials to adsorptive separation of methylalumoxane used as co-catalyst in olefin polymerization. Microporous Mesoporous Mater 44–45:557
104. Sano T, Hagimoto H, Jin J, Oumi Y, Uozumi T, Soga K (2000) Influences of methylaluminoxane separated by porous inorganic materials on the isospecific polymerization of propylene. Macromol Rapid Commun 21:1191
105. Silveira F, Petry CF, Pozebon D, Pergher SB, Detoni C, Stedile FC, dos Santos JHZ (2007) Supported metallocene on mesoporous materials. Appl Catal A Gen 333:96
106. Silveira F, Pires GP, Petry CF, Pozebon D, Stedile FC, Santos JHZ, Rigacci A (2007) Effect of the silica texture on grafting metallocene catalysts. J Mol Catal A Chem 265:167
107. Wongwaiwattanakul P, Jongsomjit B (2008) Copolymerization of ethylene/1-octene via different pore sized silica-based-supported zirconocene/dMMAO catalysts. Catal Commun 10:118
108. Bunchongturakarn S, Jongsomjit B, Praserthdam P (2008) Impact of bimodal pore MCM-41-supported zirconocene/dMMAO catalyst on copolymerization of ethylene/1-octene. Catal Commun 9:789
109. Kumkaew P, Wanke SE, Praserthdam P, Danumah C, Kaliaguine S (2003) Gas-phase ethylene polymerization using zirconocene supported on mesoporous molecular sieves. J Appl Polym Sci 87:1161
110. Kumkaew P, Wu L, Praserthdam P, Wanke SE (2003) Rates and product properties of polyethylene produced by copolymerization of 1-hexene and ethylene in the gas phase with (n-BuCp)2ZrCl2 on supports with different pore sizes. Polymer 44:4791
111. Paredes B, Grieken R v, Carrero A, Suarez I, Soares JBP (2011) Ethylene/1-Hexene copolymers produced with MAO/(nBuCp)2ZrCl2 supported on SBA-15 materials with different pore sizes. Macromol Chem Phys 212:1590
112. Tioni E, Monteil V, McKenna T (2013) Morphological interpretation of the evolution of the thermal properties of polyethylene during the fragmentation of silica supported metallocene catalysts. Macromolecules 46:335
113. Shin K, Woo E, Jeong YG, Kim C, Huh J, Kim KW (2007) Crystalline structures, melting, and crystallization of linear polyethylene in cylindrical nanopores. Macromolecules 40:6617
114. Woo E, Huh J, Jeong YG, Shin K (2007) From homogeneous to heterogeneous nucleation of chain molecules under nanoscopic cylindrical confinement. Phys Rev Lett 98:136103

Adv Polym Sci (2018) 280: 65–100
DOI: 10.1007/12_2017_8
© Springer International Publishing AG 2017
Published online: 17 May 2017

# Green Emulsion Polymerization Technology

**Yujie Zhang and Marc A. Dubé**

**Abstract** The polymer industry is dominated by the use of petroleum-based feedstock and, as a result of increased awareness, the related environmental consequences have provided the impetus for change. Emulsion polymerization is considered to be a more sustainable technique for the manufacture of polymeric materials because of its use of water as a dispersing medium. To further improve the sustainability of emulsion polymerization technology, the "12 principles of green chemistry and engineering" were used as a guideline for design of a greener process. The most obvious and effective approach is to use renewable, biobased feedstock in emulsion polymerization formulations. In addition, maximizing energy efficiency, preventing waste and pollution, and minimizing the potential for accidents also figure prominently.

**Keywords** Emulsion polymerization • Renewable feedstock • Sustainability

## Contents

Y. Zhang and M.A. Dubé (✉)
Department of Chemical and Biological Engineering, Centre for Catalysis Research and Innovation, University of Ottawa, 161 Louis Pasteur Pvt., Ottawa, ON, Canada, K1N 6N5
e-mail: Marc.Dube@uOttawa.ca

# 1  Introduction

Polymeric materials in their various forms (e.g., plastics, paints, and rubbers) play a significant role in every aspect of human life and there is no doubt that their technological impact has in many ways improved our standard of living. However, with the enormous growth of the polymer industry, non-negligible environmental consequences have surfaced. For example, synthetic polymers are not normally biologically degradable and this has led to their significant accumulation (about 22–43% of global plastics) in municipal landfill sites [1]. According to the U.S. National Institutes of Health, about 44% of seabird species are known to have ingested synthetic polymers mistakenly, and the same thing has happened to 267 marine species [2]. At the same time, many components in synthetic polymers (e.g., residual monomers, catalysts, and additives), which in most cases are toxic, can migrate into the environment and inevitably endanger the health of both wildlife and humans. In addition, there is great concern about emissions of air pollutants (e.g., carbon dioxide, methane, and volatile organic compounds; VOCs) that are involved in the production and waste management of synthetic polymers. Since the Industrial Revolution, greenhouse gas emissions have been on the increase (Fig. 1). This matter has captured world attention, and as of 24 June 2016, 177 countries including the European Union have signed the Paris Agreement, which aims to cut greenhouse gas emissions worldwide and limit global warming to below 2°C by 2030 [4].

    With growing concerns about the environmental impact of polymers, it is absolutely necessary to carry out immediate action to encourage the synthesis of more sustainable polymer products. To start with, a green synthesis pathway needs to be chosen. Emulsion polymerization is considered to be a more sustainable and

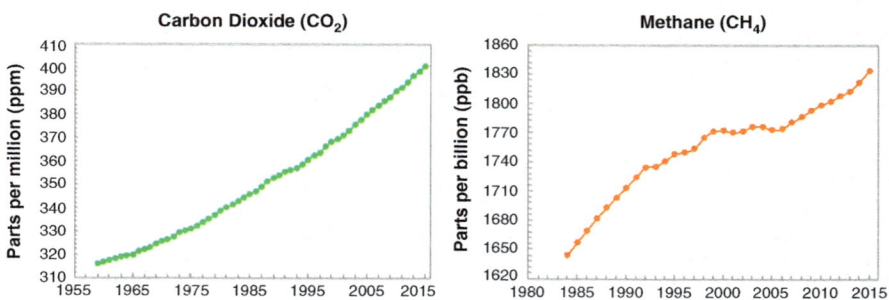

**Fig. 1**  Major greenhouse gases since the Industrial Revolution [3]

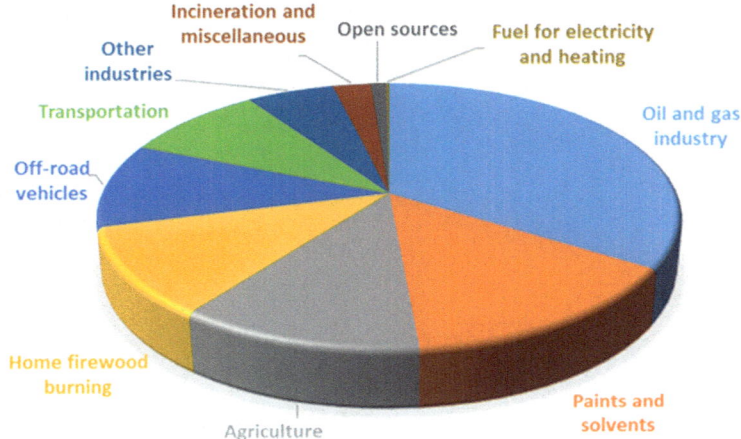

**Fig. 2** Distribution of VOC emissions in Canada by source, 2014 [5]

**Table 1** Typical emulsion polymerization formulation

| Component | Chemical | Weight (g) |
|---|---|---|
| Monomer | Butyl acrylate (BA) | 120 |
| Solvent | Deionized water | 300 |
| Initiator | Potassium persulfate (KPS) | 1 |
| Surfactant | Sodium dodecyl sulfate (SDS) | 10 |
| Buffer | Sodium carbonate | 0.5 |

environmentally friendly way to produce a wide range of polymers because water is used as a suspending medium rather than solvent. The method not only eliminates the consumption of organic solvents, which are one of the main sources of VOCs (Fig. 2) [5], but also acts as an excellent heat sink to facilitate control of the reaction temperature during synthesis. Typical emulsion polymerization formulations contain several components (e.g., monomer, initiator, surfactant, and buffer) (Table 1) that influence final polymer properties [6]. Polymers manufactured by emulsion polymerization can be found in various applications, for instance, paints and coatings, adhesives, plastics, and synthetic rubber. In 2015, the global emulsion polymer market was around $33.30 billion US dollars, and is expected to grow continuously in the near future [7].

Even though emulsion polymerization is considered to be the better environmental choice over other polymer synthesis methods, considerable environmental concerns surrounding latex polymer production remain. Thus, it is of great importance to reconsider all aspects of emulsion polymerization technology to achieve a more effective pathway towards sustainable polymer products. In order to guide us

**Fig. 3** The 12 principles of green chemistry [8]

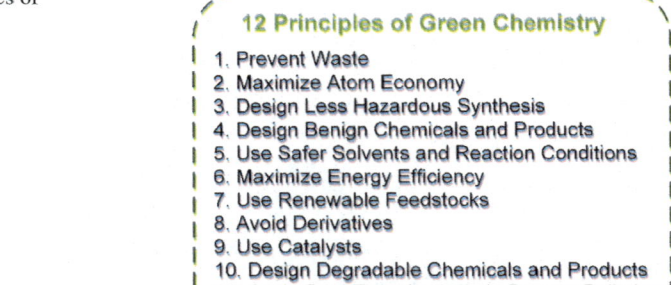

**12 Principles of Green Chemistry**

1. Prevent Waste
2. Maximize Atom Economy
3. Design Less Hazardous Synthesis
4. Design Benign Chemicals and Products
5. Use Safer Solvents and Reaction Conditions
6. Maximize Energy Efficiency
7. Use Renewable Feedstocks
8. Avoid Derivatives
9. Use Catalysts
10. Design Degradable Chemicals and Products
11. Apply Real-Time Analysis to Prevent Pollution
12. Prevent Accidents

**Fig. 4** The 12 principles of green engineering [9]

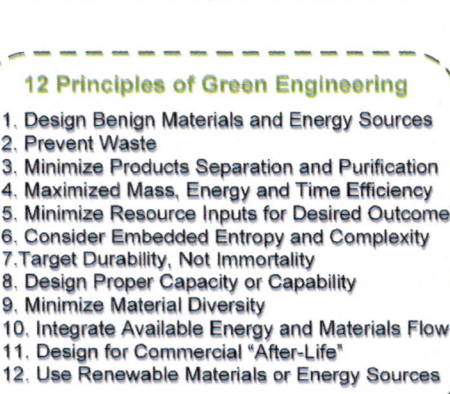

**12 Principles of Green Engineering**

1. Design Benign Materials and Energy Sources
2. Prevent Waste
3. Minimize Products Separation and Purification
4. Maximized Mass, Energy and Time Efficiency
5. Minimize Resource Inputs for Desired Outcome
6. Consider Embedded Entropy and Complexity
7. Target Durability, Not Immortality
8. Design Proper Capacity or Capability
9. Minimize Material Diversity
10. Integrate Available Energy and Materials Flow
11. Design for Commercial "After-Life"
12. Use Renewable Materials or Energy Sources

towards a greener emulsion polymerization technique, some principles can be used at the design stage. The "12 principles of green chemistry" (Fig. 3) are based on the fundamental science of chemistry to enable scientists to minimize or eliminate the environmental impact of chemical products and processes by reducing or eliminating hazardous chemicals, using renewable feedstocks, catalysts, etc. [8]. These 12 principles have also been expressed from an engineering perspective (Fig. 4) [9]. The principles have been applied by scientists and engineers in different fields of study to pursue more sustainable solutions. Recently, Dubé and Salehpour applied the 12 principles of green chemistry to polymer production technology [10].

In this chapter, the 12 principles of green chemistry and engineering will be applied to conventional emulsion polymerization techniques to design a greener path to polymer manufacture. The use of these principles implies that we need to design not only less hazardous chemicals and products, but also more sustainable processes. Some of these principles (2, 8, and 9) are already well addressed in polymerization processes in general and are not be addressed here. The use of safer solvents (principle 5) is evidently addressed in emulsion polymerization, a water-based process.

To begin with, we focus on possible renewable alternatives for some main components in the emulsion polymerization formulation. This addresses principles 3, 4, and 7 (i.e., less hazardous synthesis, designing benign chemicals and products, and using renewable feedstock), and in an indirect way, principle 10 (designing degradable products). We then turn to the development of a more sustainable emulsion polymerization from a process technology point of view: how to prevent waste (principle 1), how to maximize energy efficiency (principle 6), real-time analysis (principle 11), and how to minimize the potential for accidents (principle 12).

# 2 Design of Green Chemicals and Products

According to the 12 principles of green chemistry, to achieve green emulsion polymer products, the chemicals involved in emulsion formulations (e.g., monomers, surfactants, and initiators) should be less hazardous and derived from renewable sources. Here, we focus on emulsion polymerization formulations and review possible alternative "green" components to develop a more sustainable product. Monomer is the most plentiful ingredient in emulsion polymerizations and we therefore provide a more comprehensive review on this component. Other components such as surfactant, crosslinker, chain transfer agent, initiator, and buffer are addressed more briefly.

## 2.1 Monomer

Monomers, the main component in emulsion polymerization formulations (Table 1), typically comprise 30–60% of the total latex mass [11]. The monomers undergo free radical polymerization, which implies the presence of at least one vinyl group in the monomer structure. The principle monomer in the formulation requires a relatively low water solubility to form particles, otherwise the reaction proceeds as an aqueous solution polymerization. To achieve desired polymer properties, a multimonomer mixture (i.e., co-monomers) is commonly used in emulsion polymerization.

In general, monomers used in emulsion polymerization are obtained from fossil-based resources. Around 7% of global fossil fuel production is diverted for plastics manufacture [12]. With the uneven distribution and depletion of fossil resources, there is no guarantee of a stable price and supply chain and, as noted earlier, numerous environmental concerns arise over the use of this nonrenewable feedstock. At the same time, petroleum-derived monomers are usually toxic and pose a potential health risk to personnel exposed to these materials. For instance, exposure to acrylic monomers can lead to clinical symptoms and some of the monomers are possibly carcinogenic to humans [13–15]. To create safer products and work

**Table 2** Categorized renewable monomers

| Compound category | Monomers |
| --- | --- |
| Terpenes | Limonene, α-pinene, β-pinene, myrcene |
| Vegetable oils | Linoleic acid, linolenic acid |
| Sugars | Anhydroalditols, aldonic acids, lactones |
| Polysaccharide | Starch, cellulose[a] |
| Furan | Furfural, hydroxymethylfurfural |
| Rosin | Abietic acid, pimaric acid |

[a]Starch and cellulose are, of course, already polymers and are used frequently as fillers, surfactants, or in graft polymerization. In this context we refer to their use as macromonomers

environments, it is essential to pursue less toxic alternatives to the monomers used currently. Many renewable monomers, being derived from plants, are nontoxic; however, the origin of a material and its toxicity are not necessarily correlated.

Renewable monomers have been available for decades and are either obtained directly or derived from natural resources. Because natural resources consume $CO_2$ through photosynthesis, using natural resources as feedstock slows $CO_2$ buildup in the atmosphere. In addition, not only are renewable monomers abundant, they also offer a wide variety of building blocks that may not exist in fossil-based resources. A common refrain is that biobased polymers present poorer performance compared with those derived from petroleum-based monomers. However, by applying advanced polymer chemistry and reaction engineering techniques, one can optimize these new processes to achieve similar and sometimes better properties. Economic feasibility of these processes is also of concern. Often, raw material prices for biobased feedstock are low, but if extensive purification and derivatization are required, feedstock costs become significant. It should be made clear that despite being renewably sourced, all aspects of production of a material (e.g., cultivation, downstream modification) should be considered when assessing the "greenness" of a monomer.

Renewable monomers can be classified in various categories depending on their source (Table 2) [16]. However, only monomers containing vinyl groups and having low water solubility are appropriate for emulsion polymerization. Some renewable monomers can be applied directly in emulsion polymerization after a simple extraction or purification process (e.g., limonene) [17]; others require minor modification (e.g., conjugated linolenic acid) [18]; and others may have to be greatly modified to produce "new" monomers. In the latter case, modification costs and sustainability of the derivatization process must be balanced with the renewability of the initial feedstock. Research on the use of renewable monomers in emulsion polymerization is not extensive, therefore we provide a review of renewable monomers that have either been studied in or show promise for application in emulsion polymerization.

**Fig. 5** Monoterpene monomers

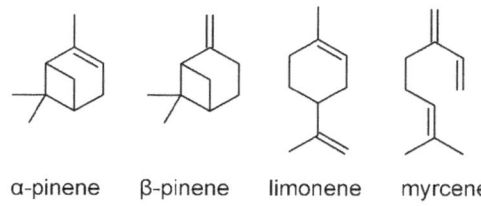

α-pinene  β-pinene  limonene  myrcene

## 2.1.1 Terpenes

Terpenes are a large group of compounds, existing mainly in plants (e.g., conifers, citrus fruits), that share the same isoprene building block [19, 20]. Terpene monomers that consist of two isoprene units with the molecular formula $C_{10}H_{16}$ are classified as monoterpenes (Fig. 5). Monoterpenes are extremely diverse as a result of their variety of molecular backbone structures, stereoisomers, and derivatives. Many monoterpenes have been used as fragrances, in foods as additives, and as green solvents. Because most monoterpenes (e.g., limonene, α-pinene) contain double bonds, they can be utilized as potential starting materials for polymer production.

Studies have shown that many monoterpenes can be polymerized via cationic initiation; nevertheless, a limited number can undergo free radical polymerization in emulsion [19, 21, 22]. The presence of allylic C–H bonds in monoterpenes makes it difficult to homopolymerize those monomers using free radical polymerization. Free radical homopolymerization of monoterpenes leads to fairly low molecular weight polymers [23, 24]. However, copolymerization offers greater opportunity for incorporation of monoterpenes in emulsion polymerization.

### Pinene

Pinene is a monoterpene with a bicyclic molecular structure (Fig. 5). α-Pinene (α-PIN) and β-pinene (β-PIN) are two isomers available in nature, and both can be obtained from conifers or other non-coniferous plants by steam distillation [20, 25]. Many pinene derivatives (e.g., 3-carene, myrcene, and limonene) can be obtained by isomerization of pinene [26].

Free radical polymerization of α-PIN and β-PIN were investigated using azobisisobutyronitrile (AIBN) as initiator at 60°C [23]. Only oligomers were produced from attempts to homopolymerize both isomers. However, when α-PIN and β-PIN were copolymerized with methyl methacrylate (MMA) and styrene (STY), relatively high molecular weight polymers were obtained (Table 3). It was shown that α-PIN is more reactive than β-PIN, which may result from the different isomers of the propagating radicals. β-PIN was successfully copolymerized with pentafluorostyrene using benzoyl peroxide (BPO) as initiator at 70°C [27]. Furthermore, reversible addition-fragmentation transfer (RAFT) radical copolymerization

**Table 3** Overall conversion and weight-average molecular weight ($M_w$) data ($T = 60°C$) [23]

| Polymer | Conversion (wt%) | $M_w$ |
|---|---|---|
| Poly(α-PIN) | 6 | 850 |
| Poly(β-PIN) | 5 | 880 |
| Poly(α-PIN-*co*-MMA) | 40 | 53,200 |
| Poly(β-PIN-*co*-MMA) | 21 | 11,600 |
| Poly(α-PIN-*co*-STY) | 7 | 25,800 |
| Poly(β-PIN-*co*-STY) | 5 | 25,300 |

of β-PIN with methyl acrylate [28], acrylonitrile [29], *n*-butyl acrylate (BA) [30], and *n*-phenylmaleimide [31] were conducted. The estimated reactivity ratios of β-PIN in all of the above cases were close to zero, which implies that β-PIN tends to react with other monomers but not with itself. All of the above work suggests that both α-PIN and β-PIN can be considered as alternatives for production of more sustainable polymeric materials via emulsion polymerization and, more specifically, emulsion copolymerization.

## Limonene

Limonene is a monocyclic terpene derived from citrus fruit oils and many other essential oils. Because limonene is a chiral monomer, there are two isomers, *d*-limonene (also written as (+)-limonene) and *l*-limonene (or (−)-limonene). Most naturally occurring limonene is *d*-limonene, which is obtained from citrus fruit oils via distillation. Limonene is a common additive in cosmetics, food, and medicines because of its pleasant orange odor [32]. Also, limonene is used extensively as a green solvent for cleaning purposes [33].

Two carbon double bonds in limonene suggest the possibility of free-radical polymerization. Similar to pinene, limonene presents some challenges for homopolymerization. Free radical copolymerization of *d*-limonene and *l*-limonene with maleic anhydride was first presented in 1994, and alternating copolymers were produced using AIBN or BPO as initiators [34]. Attempts to copolymerize limonene with STY [35], acrylonitrile [36], and vinyl acetate [37] in solution polymerization have been reported but conversions were limited to 20 wt%.

Recent investigations of the bulk copolymerization of *d*-limonene with butyl methacrylate (BMA) [17], 2-ethylhexyl acrylate (EHA) [38], and BA [39] led to the production of homogeneous copolymers with relatively high molecular weights (>100,000). Nevertheless, it was revealed that a significant degradative chain transfer mechanism manifests itself in all systems, and this leads to limited monomer conversions and lower molecular weight. Using the reactivity ratios estimated from that work, one can employ a semibatch monomer feed policy in emulsion polymerization to increase *d*-limonene incorporation [40].

Limonene (and limonene derived from pinenes) can be dehydrogenated to α-methyl styrene, which can be polymerized via a free-radical mechanism [41, 42].

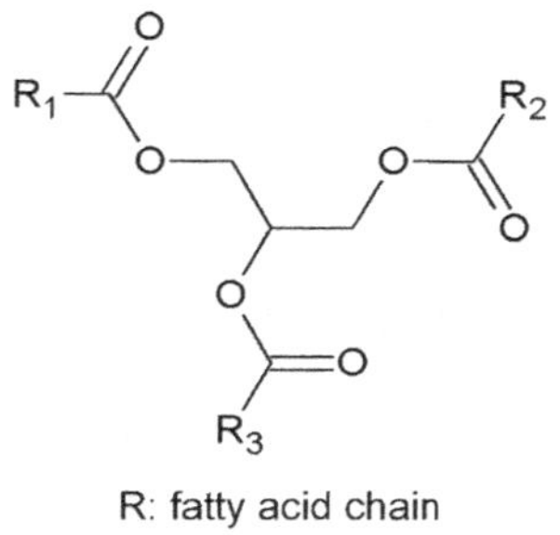

**Fig. 6** General structure of triglycerides

R: fatty acid chain

**Table 4** World production of major vegetable oils in 2016 [43]

| Oil | Annual production (million metric tons) | Main oil producer |
|---|---|---|
| Palm | 65.50 | Indonesia |
| Soybean | 53.82 | China |
| Rapeseed | 26.50 | European Union |
| Sunflower seed | 16.24 | Ukraine |
| Palm kernel | 7.66 | Indonesia |
| Peanut | 5.54 | China |
| Cottonseed | 4.50 | China |
| Coconut | 3.41 | Philippines |
| Olive | 3.01 | European Union |

Limonene-derived $\alpha$-methyl styrene has been used to produce sustainable thermoplastic elastomers with improved mechanical properties [41].

### 2.1.2 Vegetable Oils

Vegetable oils are triglycerides (Fig. 6) extracted from plants (e.g., soybean, sunflower, and peanut) via mechanical or chemical extraction. For centuries, many vegetable oils have been used in cooking, cosmetics, and personal care products. In addition, many oils have been incorporated into coatings, inks, lubricants, and wood treatment products. In 2016, the total world major vegetable oil production was 186.17 million metric tons [43]. The annual production of each major vegetable oil and the major producers are listed in Table 4. Vegetable oils have become good candidates for a renewable resource for production of polymeric materials because of their abundance, large variety and, of course, their functionality, which could lead to direct polymerization or straightforward derivatization.

The physical and chemical properties of vegetable oils vary according to their species. In general, each vegetable oil contains a mixture of triglycerides with different fatty acid chains that may have carbon chain lengths of 14–22 and 0–5 double bonds, or some other functional groups (e.g., epoxy rings, hydroxyl moieties, or ether groups) [44]. The chain length and the number and position of double bonds also have an important effect on the oil properties. Via transesterification using an alcohol, triglycerides can be converted to fatty acids (or fatty esters) and glycerol. The presence of double bonds in vegetable oils, as well as in the fatty acids (or fatty esters) derived from them, opens up possibilities for their application as renewable monomers in emulsion polymerization. Figure 7, shows some representative fatty acids with double bonds that can be found in vegetable oils.

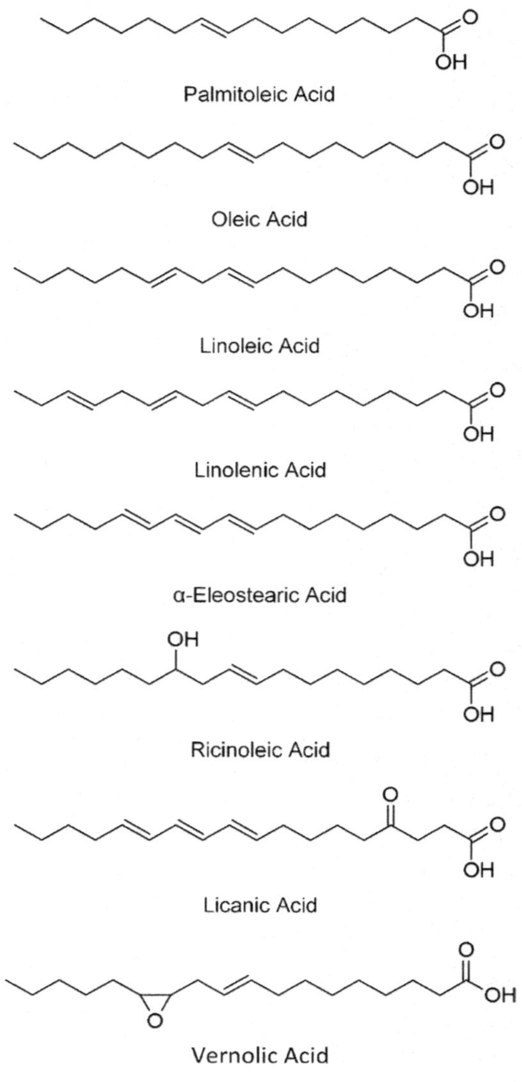
O
OH
Palmitoleic Acid
O
OH
Oleic Acid
O
OH
Linoleic Acid
O
OH
Linolenic Acid
O
OH
α-Eleostearic Acid
OH
O
OH
Ricinoleic Acid
O
O
OH
Licanic Acid
O
OH
O
Vernolic Acid

**Fig. 7** Fatty acids with double bonds

Direct Use of Vegetable Oils

Vegetable oils relatively rich in double bonds (e.g., tung oil, linseed oil, and soybean oil) can be used in polymerization directly or after slight modification (e.g., conjugation). A small amount of tung oil (0.025–0.7 wt%) was successfully copolymerized with STY using a combination of free radical initiators to obtain better molding properties [45]. More significant tung oil quantities were used in a bulk copolymerization with STY [46] and 30–50 wt% tung oil was used in a terpolymerization with STY and divinylbenzene (DVB) [47].

Highly conjugated linseed oil was reacted with STY and DVB at concentrations ranging from 30 to 70 wt% at 85–160°C.Fully cured thermosets that contain approximately    35–85%    crosslinked    materials    were    produced [48]. Terpolymerization of fully conjugated linseed oil with acrylonitrile (AN) and DVB was performed in bulk. From 30 to 75 wt% conjugated linseed oil was used in the polymerization, and thermosets with a maximum oil incorporation of 96 wt% were produced [49].


Modified Vegetable Oils

To improve vegetable oil incorporation in polymers, a more substantial modification of the vegetable oil can be performed (e.g., epoxidation, transesterification, or malination).

Macromonomers were produced via intertransesterification of castor oil and linseed    oil,    followed    by    esterification    with    acrylic    acid    [50].    The homopolymerization of these macromonomers proved difficult due to steric hindrance of the macromonomer structure. However, copolymerization with STY was successful, and copolymers with good film properties were produced. A similar approach was used to produce macromonomers by first reacting with glycerol and then transesterification with MMA; these were also copolymerized with STY [51].

Soybean oil and sunflower oil were bromoacrylated with the addition of acrylic acid (AA) and N-bromosuccinimide (Fig. 8) [52]. Bromoacrylation was conducted at room temperature with the addition of N-bromosuccinimide and acrylic acid. However, the limited yield (75% for soybean oil, 55% for sunflower oil) means that an extra step is needed to remove unreacted monomers. The bromoacrylated soybean and sunflower oils were then copolymerized with STY. Bromoacrylated soybean oil yielded rigid polymer, whereas the sunflower oil yielded soft polymer. It should also be noted that N-bromosuccinimide is a chemical hazard and this reduces the "greenness" of the approach.

Acrylated-epoxidized soybean oil (AESO) (Fig. 9) is a well-studied modified vegetable oil and can be produced in two steps: epoxidation of soybean oil, and then acrylation using acrylic acid [53]. The level of acrylation greatly influences the mechanical properties of AESO. Photopolymerization of AESO resulted in high yield of crosslinked polymer in a relatively short time frame [54]. AESO has been copolymerized    with    BMA    to    manufacture    electrically    conductive    polymer

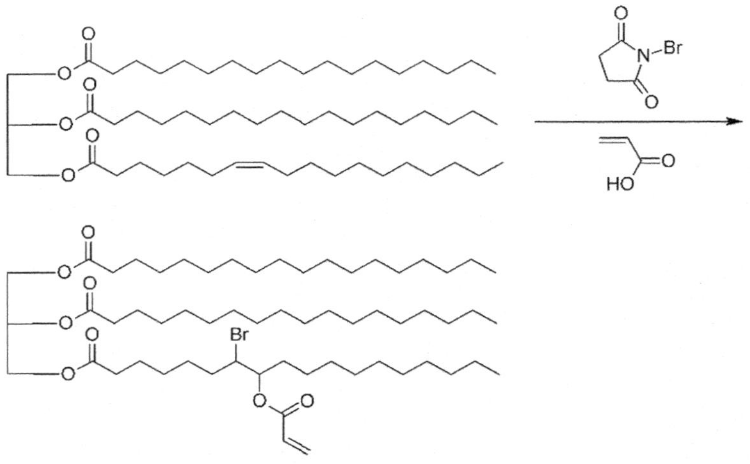

**Fig. 8** Synthesis of bromoacrylated triglycerides

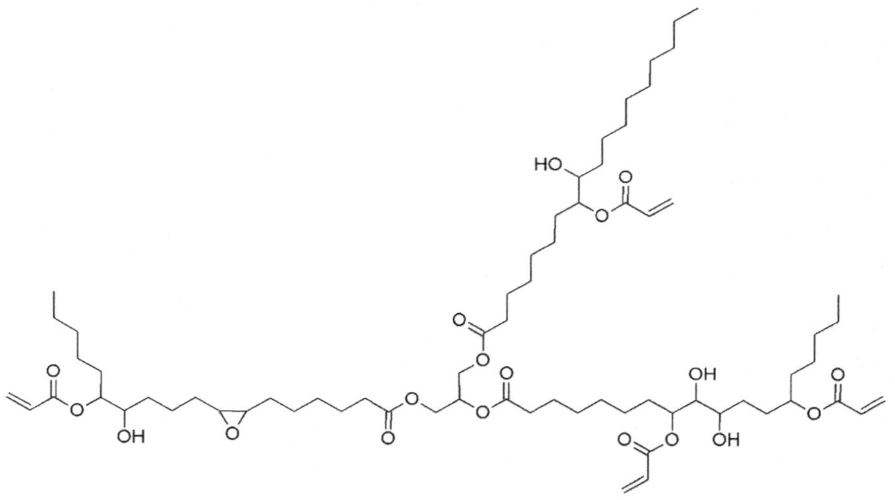

**Fig. 9** Example structure of AESO

composites [53] and has also been used in pressure-sensitive adhesive formulations [55].

Soybean oil and castor oil were also modified by the alcoholysis reaction with polyol followed by malination with maleic anhydride [56, 57]. The resulting macromonomers were copolymerized with STY to yield a thermoset polymer.

## Monomers Derived from Vegetable Oils

Beyond the direct use or modification of vegetable oils, there are many examples of monomer production (e.g., glycerol, fatty acids, and fatty esters) using transesterification or other similar reaction. These monomers and their derivatives are yet another alternative for application in emulsion polymerization.

### Fatty Acids and Their Derivatives

As discussed above, fatty acids derived from vegetable oils normally contain long carbon chains and double bonds or other functional groups. Those with double bonds (Fig. 7) can be used directly in emulsion polymerization. For example, conjugated linoleic acid (CLA) was successfully used in bulk and emulsion polymerization [58, 59]. CLA was used as a bulk co- and terpolymer with BA and STY, and in emulsion as a CLA/BA/STY terpolymer. Application of emulsion terpolymer latex film as a pressure-sensitive adhesive was demonstrated, with up to 30 wt% CLA content in the latex.

Saturated fatty acids can be used in emulsion after a derivatization such as hydrogenation to fatty alcohols [60]. Further reaction with AA (which can also be derived from natural resources) can yield biobased acrylic monomers [61–63]. 2-Octonal derived from ricinoleic acid, a product of castor oil transesterification, was used to produce 2-octyl methacrylate for latex-based adhesives [61].

### Glycerol and Its Derivatives

Glycerol is the backbone of triglycerides and can be obtained by hydrolysis or transesterification of triglycerides from plant and animal resources. As the byproduct of biodiesel, the production of glycerol is expected to exceed market demand sixfold by 2020 [64]. Glycerol has been used either in monomer or polymer form in the food, cosmetics, and pharmaceuticals industries [65].

Glycerol (Fig. 10) is a polyol compound and cannot be used directly as a monomer in emulsion. However, its derivatives (Fig. 10) obtained by modifying the alcohol groups in its structure can be used in free radical polymerization [66, 67]. Normally, these can be modified via acrylation, alkylation, chlorination, dehydration, and so on [53]. Free radical and living polymerization of glycerol-based monomers have been reviewed recently [68]. Some examples include glycerol dimethacrylate (Fig. 10) copolymerized with STY and 4-vinyl-pyrrole [69, 70]; and solketal methacrylate (Fig. 10) copolymerized with tert-butyl methacrylate, 2-(N,N-dimethylamino) ethyl methacrylate, and 2-bromoethyl methacrylate [71]. In addition, glycerol can be converted to allyl alcohol, which can further undergo free-radical polymerization [66, 72].

There is no doubt that various derivatized glycerol monomers can be developed and used in emulsion polymerization. Nevertheless, the modification of glycerol could come at significant economic and environmental cost.

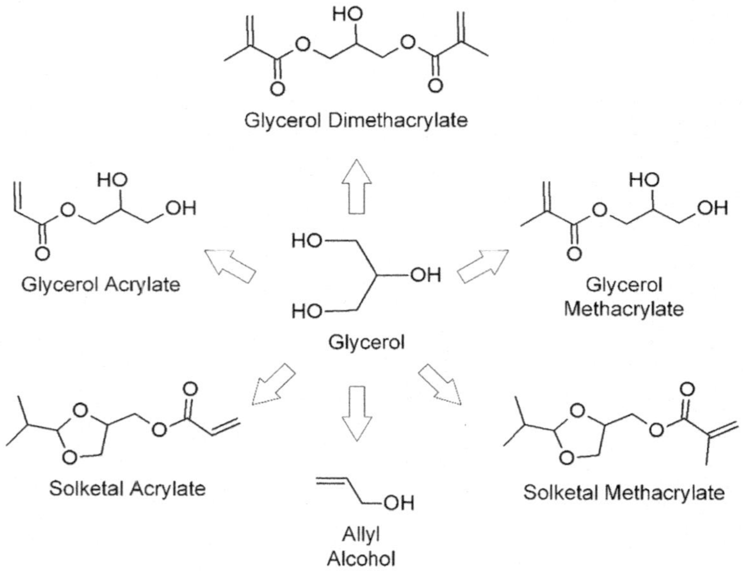

Fig. 10 Glycerol and its derivatives

### 2.1.3 Sugar-Based Monomers

Sugars are carbohydrates that exist in the tissues of many plants (e.g., grapes, banana, sweet potato, and yam) and are normally extracted from sugar beet and sugar cane. In 2016, global sugar production was 169 million metric tons [43]. Sugars are abundant and possess an enormous variety of structures with multifunctional groups [73]. Nevertheless, their multifunctionality generally needs to be reduced to avoid undesired byproducts. Polymerization of sugar-based monomers often yields atactic polymers, but greater stereoregularity can be achieved [74].

Generally, sugar-based monomers are used to produce polyesters [75–77], polycarbonates [78, 79], and polyurethanes [80, 81] via step-growth polymerization. However, some of those monomers can be modified for application in emulsion polymerization.

AA is a monomer often added to emulsion formulations to improve latex stability and film properties [82]. AA is produced commercially from petroleum-sourced propylene. However, it is possible to obtain AA through fermentation of sugars via several pathways (Fig. 11). One method involves first obtaining lactic acid by fermentation and then dehydrating the lactic acid to produce AA [83, 84]. In 2011, the Dow Chemical Company and OPX Biotechnologies Inc., announced the development of a process using fermentation of sugar-based 3-hydroxypropionic acid [85, 86]. A more recent report involves the metathesis transformation of sugar-based fumaric acid [87].

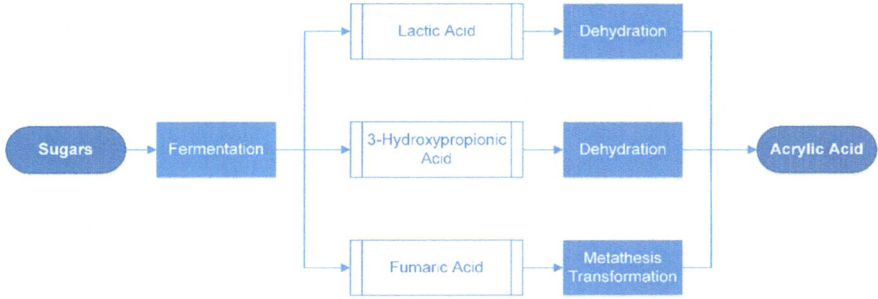

**Fig. 11**  Pathways to sugar-based AA

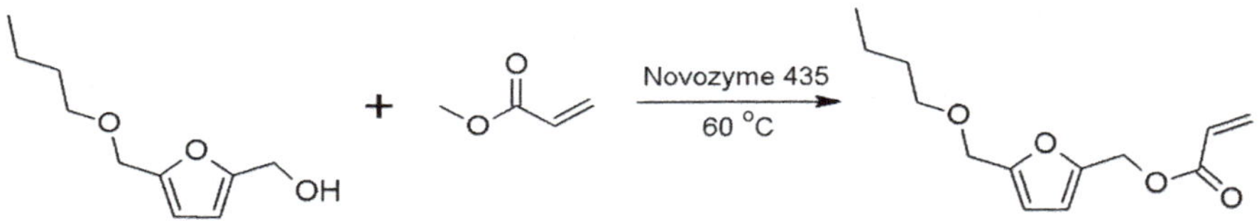

**Fig. 12**  Reaction to produce 5-butoxymethylfurfuryl acrylate

It is also possible to derive sugar-based acrylic and methacrylic monomers that can be applied in emulsion polymerization. *n*-Butanol is an important building block that can be derived from the fermentation of sugars and starches [88]. Although it cannot be directly used in emulsion polymerization, it can be used to produce biobased BA via acrylation using AA [60]. Biobased MMA can be produced via fermentation of sugar cane [89], or using sugar-based isobutyric acid [90]. Other acrylic and methacrylic monomers have been obtained using various sugar-based building blocks (e.g., isosorbide, hydroxymethylfurfural) [91]. For instance, a sugar-based monomer 5-butoxymethyl furfuryl alcohol can be reacted with methyl acrylate using lipase B enzyme (Nuvozyme 435) as catalyst to produce 5-butoxymethylfurfuryl acrylate (Fig. 12), which can be later polymerized in emulsion.

Another technology developed by EcoSynthetix Inc. introduced sugar-based vinyl monomers that can be used in emulsions [92, 93]. In this technology, an aldose sugar is converted to alkyl polyglycosides (APG), and then APG is reacted with maleic anhydride to produce sugar-based vinyl monomers. The sugar-based vinyl monomers were successfully applied in emulsion polymerization formulations with monomers such as BA, MMA, and vinyl acetate. This invention was targeted for application in coatings, adhesives, and toners in paper and paperboard products.

### 2.1.4 Other Biosourced Monomers

The use of terpenes, vegetable oils, sugars and their derivatives for emulsion polymerization is an emerging field of research. As shown in Table 2, a great many other possibilities exist. In all cases, however, the cost and compositional variability of the source material and the cost of derivation, if necessary, must be considered in selection of these materials. At the same time, the replacement of a petroleum-based monomer in the reaction formulation will most certainly lead to significant process modification to achieve similar or improved properties.

## 2.2 Surfactant

Surfactant (or stabilizer) is an essential component in emulsion polymerization to achieve stable latex particles and lower latex viscosity. Typical surfactants contain both a hydrophilic group (polar head group) and a hydrophobic group (alkyl chain) in their structure. They can be physically absorbed on the particle surface or chemically incorporated onto the surface. In general, there are three types of surfactants: electrostatic surfactants (ionic surfactants) (e.g., sodium dodecyl sulfate), steric surfactants (e.g., poly(ethylene oxide) nonylphenyl ether), and electrosteric surfactants, which provide both electrostatic and steric mechanisms [94–96]. Each surfactant can be used separately or as a mixture, although anionic surfactants are used most frequently.

The surfactant type dictates the amount of surfactant required in the emulsion formulation. For example, anionic surfactant is normally used at a concentration of 0.2–3 wt% in water, whereas nonionic surfactant is used in amounts of 2–10 wt% [11]. Commonly used surfactants are considered toxic to marine organisms and may result in bioaccumulation [97, 98]. Because of their relatively high cost and effect on latex application properties, minimization of surfactant concentration in emulsion formulations is highly desired.

Surfactants are produced from both petroleum and natural resources, but mostly from the former. Nevertheless, driven by environmental concerns, there is a growing trend towards the development and use of renewable and less hazardous surfactants [97]. Surfactants that are entirely derived from natural resources are commercially available; the most popular of these are shown in Table 5.

### 2.2.1 Alkyl Polyglycosides

Alkyl polyglycosides (APGs) (Fig. 13) are manufactured completely from natural resources and have been under development for over 30 years. The hydrophobic parts of APGs are fatty alcohols with chain lengths of 8–16, derived from vegetable oils (e.g., coconut oil, palm oil, and rapeseed oil), whereas the hydrophilic parts are

**Table 5** Some commercially available renewable surfactants

| Category | Supplier | Application | Global production (ton/year)[a] |
|---|---|---|---|
| Alkyl polyglycosides | Dow Chemical Company, BASF, DuPont, E. I. du Pont de Nemours and Company, Cognis, AkzoNobel, Clariant, Shanghai Fine Chemicals and LG Household & Health Care | Household detergents, personal care, cosmetics, industrial cleaners, and agricultural chemicals | >90,000 |
| Sorbitan esters | Dupont, BASF, SABO, SEPPIC, Kao, Vantage, Clariant, Lonza, Croda | Food, pharmaceutical, and cosmetic products | 25,000 |
| Sucrose esters | BASF, Evonik Industries, P&G Chemicals, Croda International Plc, Sisterna B.V., Mitsubishi Chemical Holdings Corporation, Dai-Ichi Kogyo | Food and beverage additives, personal care, detergents, and cleansers | <10, 000 |

[a]References are cited in the text

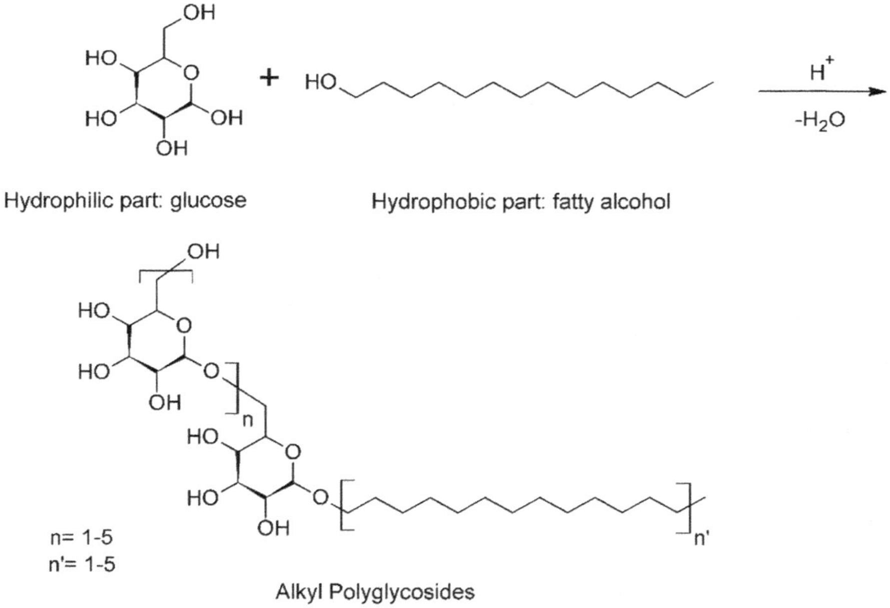

Hydrophilic part: glucose          Hydrophobic part: fatty alcohol

n= 1-5
n'= 1-5

Alkyl Polyglycosides

**Fig. 13** Synthesis of alkyl polyglycosides

glucose-derived from starch, potatoes, sugars, etc. In general, APGs are mixtures of isomers, and are characterized according to alkyl chain length and the degree of oligomerization ([99, 100]). Commercially available APGs have a degree of oligomerization ranging from 1.3 to 1.6.

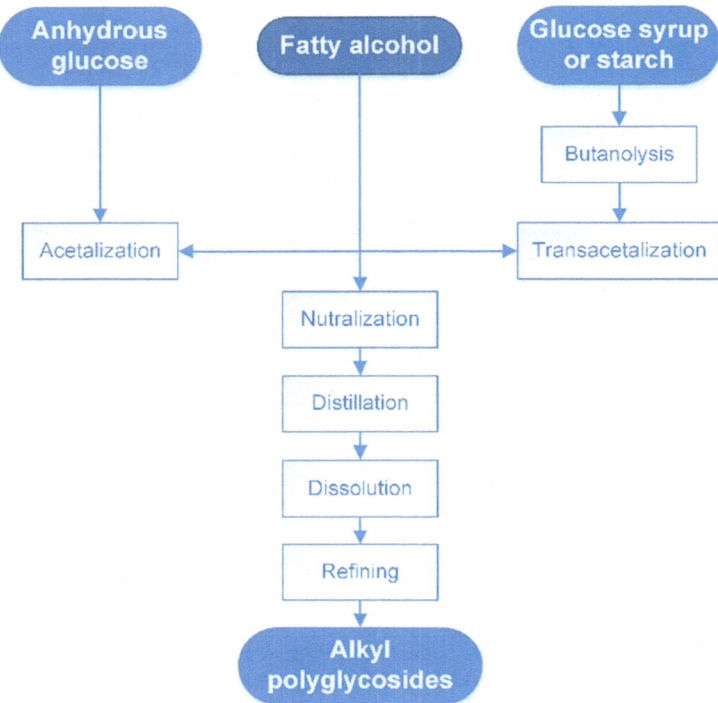

**Fig. 14** Production of alkyl polyglycosides

APGs are supplied by many companies [101, 102], with a global production of over 90,000 tons (Table 5) [103]. Industrial production of APGs normally includes several stages (Fig. 14). The first stage is acetalization of glucose with fatty alcohols. The acetylization process differs depending on the source of glucose. For anhydrous glucose, direct acetalization can be applied, whereas for glucose syrup and starch, butanolysis is conducted prior to transacetalization with fatty alcohols. A neutralization stage follows to terminate the acetalization reaction. Because excess fatty alcohol is needed for acetalization, a distillation stage is used to remove and recycle excess fatty alcohols. Additional refining (e.g., bleaching, stabilization) is sometimes necessary. The APG production process is solvent-free and has high yields and low emissions, which makes it industrially and environmentally favorable.

Beyond the favorable nature of the APG production process, APGs are not toxic to humans and are compatible with skin and eyes [104]. APGs have been applied as household detergents, personal care products, and cosmetics, to name a few [99].

APGs have been used as nonionic surfactants in emulsion polymerizations. APGs with different degrees of oligomerization (1.6–4.0) and alkyl chain lengths (8–14) were tested in the semibatch emulsion copolymerization of vinyl acetate and BA [105]. The degree of oligomerization and alkyl chain length of the APGs affected latex particle stability. As with typical nonionic surfactants, with increasing APG concentration, latex particle size decreased while particle stability increased. However, this only applied up to a certain APG concentration, beyond which destabilization occurred. APG was also used in BA/MMA-seeded emulsion polymerization as a replacement for nonylphenol ethoxylates (NPEOs) [106]. Similar effects of APG concentration on particle size and latex stability were shown. In addition, increasing APG concentration was shown to decrease the water resistance of the latex film.

Mixtures of APGs with other green nonionic or anionic surfactants have also been studied [107, 108]. APGs with alkyl chain lengths between 9 and 11, and degree of oligomerization of 1.6 were used. APGs (4–40 wt%) were mixed with fatty alcohol ethoxylate (20–76 wt%) as a surfactant mixture. As a result, much less surfactant was needed to achieve a stable latex with excellent mechanical properties [107]. Mixtures of APG and a traditional surfactant were used in the production of a wide range of emulsion polymers (e.g., BA, MMA, and EHA) and improved the block resistance of the latex film [108].

## 2.2.2  Sorbitan Esters

Sorbitan esters are nonionic surfactants made from sorbitol (hydrophilic part, derived from glucose) and fatty acids (hydrophobic part, derived from fat or vegetable oils) (Fig. 15). Sorbitan esters are biodegradable and safe to use on humans; they have been used in food, cosmetics, and pharmaceutical products. Sorbitan esters are industrially manufactured via two approaches. The first approach involves dehydration of sorbitol to produce sorbitan, followed by transesterification of sorbitan with fatty acids [109, 110]. In the other approach, sorbitol is directly esterified with fatty acids using acid or base catalysts at relatively high temperatures (200–250 °C) [111, 112]. Both methods yield mixtures of

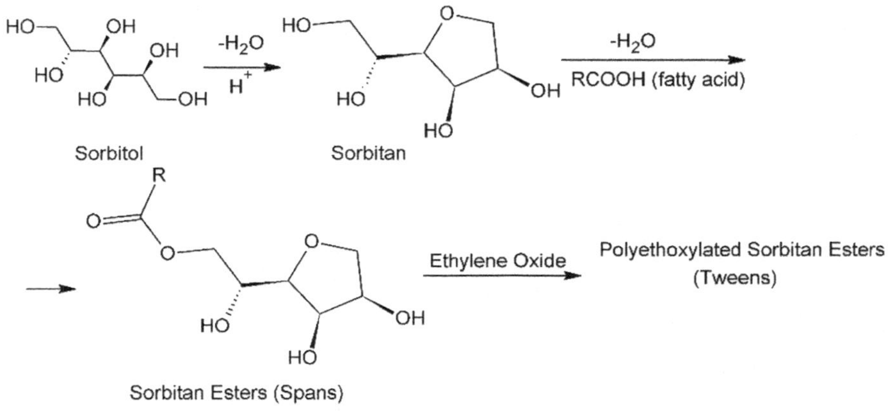

**Fig. 15** Synthesis of sorbitan esters

sorbitan fatty acid esters (commercially named "Span") with different degrees of esterification. The chain length and amount of fatty acid used, as well as the degree of esterification, influences the hydrophilic/lipophilic balance (HLB) (ranging from 1 to 8). Sorbitan fatty acid esters can be further reacted with ethylene oxide to produce polyethoxylated sorbitan esters (commercially named "Tween") with HLB values ranging from 10 to 17 [113]. The main suppliers of sorbitan esters report an annual global production of over 25,000 tons (see Table 5) [101, 103].

Spans and Tweens of different HLB values have been successfully applied in emulsion polymerization. Spans are efficient surfactants used in water-in-oil high internal phase emulsion systems [114, 115]. Tweens are generally applied in oil-in-water emulsion systems and are more commonly used in conventional emulsion polymerization [116–118]. Spans and Tweens can be used together or combined with other traditional surfactants at different ratios to obtain certain HLB values [119–122]. In general, a combination of Spans and Tweens with low and high HLB values, respectively, may work better than a single surfactant. An appropriate HLB value is required to achieve latex stability and this value can be obtained through experiment [121]. Spans and Tweens with unsaturated alkyl chains tend to provide better stability for the emulsion polymerization of unsaturated monomers [123].

### 2.2.3 Sucrose Esters

Sucrose is a carbohydrate existing in many plants (e.g., sugar cane, beet). By attaching fatty acid chains derived from vegetable oils to sucrose (Fig. 16), sucrose esters can be produced. Sucrose esters are, thus, nonionic surfactants derived completely from natural resources. In general, the production of sucrose esters poses challenges as a result of the high functionality of sucrose and its high sensitivity to temperature. Sucrose esters can be produced via transesterification of sucrose with fatty acid methyl esters (Fig. 16), or via esterification with fatty acids [124–126]. Both pathways yield mixtures of sucrose esters with various degrees of esterification, and offer sucrose esters with a broad range of HLB values. Sucrose esters have been used as food and beverage additives, and can be found in personal care products, detergents, and cleansers. In 2015, the sucrose esters market was estimated at 55.7 million US dollars and is expected to reach 74.6 million US dollars by 2020 [127]. The major sucrose ester manufacturers are listed in Table 5 [102, 127].

It is possible to use sucrose esters in emulsion polymerization, although there are limited publications on the subject. Sucrose esters were successfully used in a patent to make pH-neutral pressure-sensitive adhesives [128]. Sucrose esters, at 2–10 wt% of total polymer weight, were added as nonionic surfactants to stabilize nonionic monomers in emulsion formulations. HLB values ranging from 12 to 18 gave the best performance. Sucrose esters have also been studied in water-in-oil emulsions to produce water-absorbent resins [129]. Generally, to stabilize water-soluble unsaturated monomers (e.g., AA, sodium acrylate), 0.5–10 wt% sucrose esters relative to monomer weight are needed, and sucrose esters with HLB values between 2 and 6 are preferred.

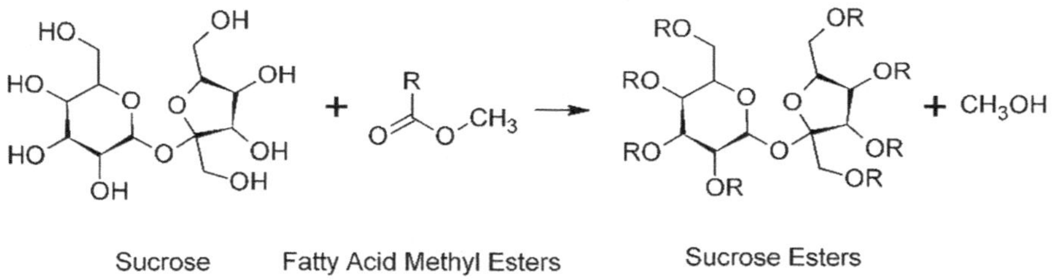

Fig. 16 Synthesis of sucrose esters with fatty acid methyl esters

## 2.3  Other Components

Compared with monomers and surfactants, relatively smaller amounts of other components (e.g., crosslinker, chain transfer agent, initiator, and buffer) are used in emulsion formulations (Table 1). Typically, these components are mostly consumed during the reaction and, thus, their environmental impact is limited. Nonetheless, replacement of some of these components with less-hazardous and nature-based chemicals is an area of active interest.

Some crosslinkers derived from renewable resources have been used in polymer synthesis to replace conventional crosslinkers [130–132]. Castor oil was used as a trifunctional crosslinker to produce polyurethane elastomers and the effects of castor oil concentration on morphology were investigated [132]. In soybean oil-based coatings, gallic acid (Fig. 17), which can be extracted from gallnuts, oak bark, tea leaves, and other plants, was used [131]. Another nature-based crosslinker 3-hydroxy-*N*,*N*-bis(2-hydroxyethyl)butanamide (HBHBA) (Fig. 17) was tested in the production of polyurethane foams as a replacement for the conventional crosslinker diethanolamine and offered better mechanical properties [130]. Biobased materials containing vinyl groups, which may be suitable for crosslinking, are generally less reactive. Pripol™ is an example of a commercial renewable crosslinker. Pripol 1009 (di-acid) and Pripol 1040 (tri-acid) are derived from vegetable oils and both have been used to produce thermosetting resins with epoxidized linseed oil [133, 134].

α-Pinene and *d*-limonene have shown potential as renewable chain transfer agents [17, 135, 136]. Investigations on the copolymerization of *d*-limonene revealed its dramatic degradative chain transfer effects on conversion, copolymer composition, and molecular weight, which makes it a renewable alternative for use as chain transfer agent [17].

Initiators not only increase polymerization rate, but are also used as "chasers" to consume residual monomer [137]. Although relatively small amounts of initiators are used, in some cases their toxicity is non-negligible [138]. To eliminate the environmental effects of initiators, either less-hazardous or highly efficient initiators should be chosen. Alternative initiation methods can also be used (discussed in Sect. 3.2.2). Buffers (e.g., sodium bicarbonate) used in emulsion polymerization are normally not of concern.

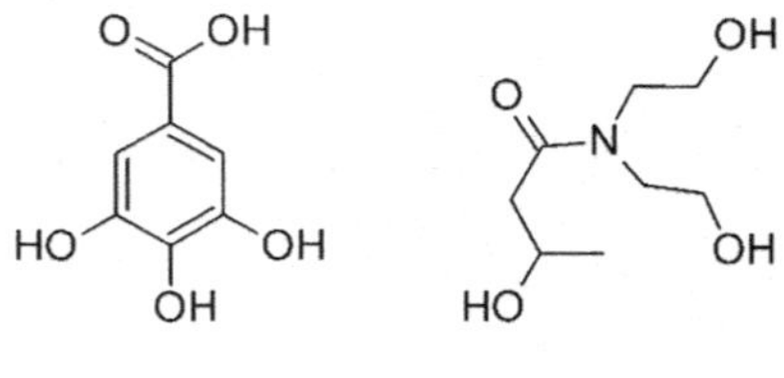

**Fig. 17** Nature-based
crosslinkers

Gallic Acid                    HBHBA

The often-present multifunctionality of renewable monomers can be explored to develop their potential as crosslinkers, chain transfer agents, etc. Furthermore, transition from petroleum-based monomers to renewable alternatives may open up possibilities for the application of other less-hazardous materials because of the additional functionality often present in biobased monomers.

# 3 Design of a Green Emulsion Polymerization Process

The use of less hazardous and renewable alternatives is an obvious approach towards achieving a "greener" emulsion polymerization. However, the sustainability or "greenness" of a process goes well beyond replacement of toxic and/or nonrenewable materials. Upon further inspection of the 12 principles of green chemistry and engineering [10], some important engineering issues in emulsion polymerization emerge. These include how to prevent waste, how to maximize energy efficiency, and how to minimize the potential for accidents; which are explored in the rest of this section. Nevertheless, other principles should still be considered during process design, and other engineering tools (e.g., life cycle assessment) can be applied to evaluate the environmental impact of the entire process [139].

## 3.1 Prevent Waste

The primary sources of waste generated in emulsion polymerization are residual monomers and production material that does not meet the specified requirements (off-spec material). To remove residual monomers, various techniques can be employed, which can be classified as chemical methods or physical methods [140]. Chemical methods are typically geared to further reaction of unreacted monomers. This can be achieved by increasing the temperature [141], adding extra initiator towards the end of the reaction [142], or adding reactive monomers [143]. Physical methods include distillation [144], supercritical devolatilization [145], spray-drying [146], and stripping [147]. These techniques can be used alone or in combination depending on final product specifications. Cost and impact on final product properties should also be considered. Off-spec material can result from impurities in the monomer and other components, inefficient mixing, or inconsistent heat transfer. Some impurities can be carefully avoided by purification of chemicals, while others can be prevented by regular maintenance and calibration [65]. Process control techniques [148] and some real-time analysis tools [149] can also be used to prevent off-spec material. In any case, one can make process operating and reaction formulation choices at the design stage to minimize the amount of residual monomer.

## 3.2 Maximize Energy Efficiency

In general, polymerizations are performed under isothermal conditions to obtain consistent polymer properties [11]. As polymerization reactions are extremely exothermic, a great deal of energy is consumed in controlling the reaction temperature. In emulsion polymerization, the aqueous medium acts as a heat sink to more easily control polymerization temperature. In large-scale reactors, challenges in heat transfer and as a result of fouling may still be present. To maximize energy efficiency and make better use of the heat produced during polymerization, one can adopt approaches such as adiabatic polymerization or alternative initiation techniques.

### 3.2.1 Adiabatic Polymerization

Adiabatic polymerization is a polymerization process that involves no external heating or cooling system or, in some cases, only involves heating at the initial stages of the process [150, 151]. This technique employs the heat generated during polymerization and is often applied to fast reactions; as a result, both reaction time and energy costs are reduced. Normally, initiators requiring relatively low temperatures (e.g., redox initiators) are used. By controlling the initiator and chain transfer agent concentrations, polymers with a range of desired properties can be achieved [150, 152].

Adiabatic polymerizations have been successfully applied to bulk and solution free radical polymerization, and thorough thermodynamic and kinetic studies have been conducted [150, 153–156]. Very recently, adiabatic emulsion polymerization was investigated using a reaction calorimeter [157]. Redox initiators (hydrogen peroxide and ascorbic acid) were used at ambient temperatures to polymerize BMA, a monomer with a relatively high reaction rate. Both batch and semibatch processes were studied to investigate the influence of adiabatic conditions on the nucleation stage. Compared with conventional emulsion polymerization conditions, a shorter reaction time and less energy were needed.

### 3.2.2 Initiation Methods

Thermal initiators are generally used in emulsion polymerization; however, some other initiation methods are available. Photo-induced emulsion polymerizations have been studied and are considered an energy-efficient method with low risk of latex destabilization [158–160]. The polymerization rate and polymer molecular weight can be controlled by changing the light intensity and irradiation time [158]. Microwave irradiation is another initiation method that can be used in emulsion polymerization to provide higher polymerization rates, higher yields, and a shorter reaction time than traditional heating [161–163]. Ultrasonication has also been studied [164–168]. Ultrasound power, pulse ratio, and probe diameter were shown to influence polymerization yields and polymer properties [167]. Ultrasound irradiation can be used in large scale latex production and shows more promise than some other methods [164, 165].

It is worth mentioning that the initiation methods mentioned above are normally used at room temperature, but one can combine these methods with adiabatic polymerization to achieve even higher energy efficiencies.

## 3.3 Apply Real-Time Analysis to Prevent Pollution

To better tune polymer product properties, minimize byproduct formation, and prevent pollution, real-time analysis tools can be used in a feedback control loop. There are various real-time analysis tools available (e.g., calorimetry, infrared spectroscopy and Raman spectroscopy), and these have been widely used in polymer production to monitor polymerization parameters (e.g., temperature, pH, and composition) [149]. In general, real-time analysis tools are cost-effective, noninvasive and environmentally friendly. Development of real-time analysis tools in emulsion polymerization is relatively slow and difficult because of the thermodynamic instability of latex particles [169]. However, there are some tools that have been successfully developed and employed in emulsion systems (e.g., calorimetry, infrared spectroscopy). Some other tools are still under development, for example, particle size analysis is of particular interest in emulsion polymerization, but reliable real-time analysis is still not available.

### 3.3.1 Calorimetry

Reaction calorimetry has been widely used to monitor polymerization by tracking the heat generated during the reaction. Heat generation is closely correlated to monomer conversion and polymerization kinetics, which makes reaction calorimetry a useful tool for predicting conversion, polymerization rate, etc. [149]. Reaction calorimetry is ideal for on-line polymerization monitoring because it is noninvasive, robust, and fast [170].

There are three types of reaction calorimetry: heat-flow calorimetry, heat-balance calorimetry, and power-compensation calorimetry [149, 171]. Heat-flow calorimetry measures the heat flow between the reactor and cooling jacket. It is of great sensitivity and, thus, is more applicable to laboratory-scale reactors as a result of the significant temperature difference between the reaction mixture and cooling jacket [172]. Heat-balance calorimetry performs energy balances based on the heat transfer of the cooling fluid and is less sensitive to temperature changes resulting from the slow flow rates of cooling fluids [149]. Power-compensation calorimetry uses a heater to achieve a constant temperature by varying the power. This technique is not commonly used in the polymer industry as cooling is not available; however, it can be effective at high pressure conditions [173, 174].

Heat-flow calorimetry has been extensively used in batch and semibatch emulsion polymerization as an on-line analysis tool [170, 175–178]. With the data collected using calorimetry, it is possible to estimate monomer conversion and copolymer composition and compare the results with data obtained by gravimetry and other conventional methods. Heat-flow calorimetry has also been used to monitor polymerization kinetics [170, 179], molecular weight [180], and nucleation [181–183]. This technique can be very helpful in adiabatic polymerization to measure the heat generation during polymerization and maximize energy efficiency [152].

### 3.3.2 Infrared Spectroscopy

Infrared (IR) spectroscopy has been developed for use in emulsion polymerization to provide information on molecular structure and polymerization kinetics [149]. IR bands have three regions: near- (wave number from 14,000 to 400 $cm^{-1}$), mid- (4,000–400 $cm^{-1}$), and far-IR (400–10 $cm^{-1}$). Near-IR can excite overtone or harmonic vibrations, whereas mid-IR can detect fundamental molecular vibrations.

Near-IR was successfully used in-line and in situ to monitor conversion and molecular weight in emulsion polymerization; however, numerous challenges were encountered [184]. Attenuated total reflectance (ATR)-Fourier transform infrared (FTIR) spectroscopy is more commonly used to monitor emulsion polymerization. In-line ATR-FTIR can detect residual monomers and estimate monomer conversion and copolymer composition in emulsion polymerization [185–188]. Results obtained via ATR-FTIR often correlate well with those determined by conventional methods.

## 3.4   Prevent Accidents

In emulsion polymerization, the aqueous medium often prevents thermal runaway; however, potential for accidents still exists. Accidents can occur during feedstock transport and storage, polymer synthesis, or when dealing with final products [10]. As mentioned above, less hazardous and renewable chemicals are far less likely to lead to hazardous situations. For example, when handling and transporting biobased chemicals, personal protective gear (although always necessary) is less crucial in the face of low or nonexistent material toxicity. During the polymerization process, the reaction temperature and pressure should be closely monitored; process control methods or equipment can be implemented in the process. For instance, early warning detection systems can be used [189]. Reactor fouling, which is an important concern for emulsion polymerizations, could change heat transfer characteristics over time. Mixing also plays an important role in heat transfer and fouling, so efficient mixing tools should be chosen to maintain constant temperature. Regular equipment maintenance and inspection should be performed to prevent fouling and detect any equipment failure (failure in cooling systems or feed pumps) that may cause accidents.

Concerns regarding final products are mainly in the event of fire, in which case hazardous chemicals may be released to the environment. This is typically prevented by the use of fire retardants or oxygen scavengers [190]. The use of biobased feedstock may mitigate the need for such additives.

## 4   Conclusion

Emulsion polymerization is often touted as a sustainable technique for the manufacture of polymer materials. Although this is true, considerable efforts are still required to minimize the environmental impact of the process and its products. The 12 principles of green chemistry and engineering establish a framework for designing green products and processes. All of the principles should be taken into consideration when designing a green product or process. Here, we have discussed selected principles based on their applicability to emulsion polymerization technology. In the design of green chemicals and products, the use of renewable feedstocks (principle 6 in the 12 principles of green chemistry) to replace their petroleum-based counterparts is recommended. Various renewable monomers and their potential for use in emulsion polymerization have been reviewed. The use of renewable feedstocks is probably the most effective means of achieving a green emulsion polymerization. Application of other principles, such as maximizing energy efficiency, using real-time analysis to prevent pollution, preventing waste and accidents, may also have a significant impact on moving towards green emulsion polymerization.

In all cases, the 12 principles of green chemistry and engineering should be actively considered when designing emulsion polymerization processes. Inevitably, there will be trade-offs in the application of different principles, but an evaluation tool such as life cycle assessment, can greatly assist in making the most sustainable choices. In other words, the consideration of producing emulsion-based polymers should lead to a scenario of least overall impact on the environment. The 12 principles of green chemistry and engineering are an appropriate guide along the way.

# References

1. Worldwatch Institute (2015) Global Plastic Production Rises, Recycling Lags. http://www.worldwatch.org/global-plastic-production-rises-recycling-lags-0. Accessed 19 Jul 2016
2. Laist DW (1997) Impacts of marine debris: entanglement of marine life in marine debris including a comprehensive list of species with entanglement and ingestion records. In: Coe JM, Rogers DB (eds) Marine debris. Springer, New York, pp. 99–139
3. US Department of Commerce (2014) The NOAA Annual Greehouse Gas Index (AGGI). http://www.esrl.noaa.gov/gmd/aggi/. Accessed 4 Jul 2016
4. United Nations (2016) The Paris Agreement. http://unfccc.int/paris_agreement/items/9485.php. Accessed 8 Oct 2016
5. Government of Canada (2016) Environment and climate change Canada. Environmental indicators – air pollutant emissions https://www.ec.gc.ca/indicateurs-indicators/default.asp?lang=en&n=E79F4C12-1. Accessed 28 Jun 2016
6. Jovanović R, Dubé MA (2004) Emulsion-based pressure-sensitive adhesives: a review. J Macromol Sci Part C 44:1–51. doi:10.1081/MC-120027933
7. Grand View Research (2016) Emulsion polymer market size and share (industry report 2022) http://www.grandviewresearch.com/industry-analysis/emulsion-polymer-market. Accessed 4 Jul 2016
8. Anastas PT (1998) Green chemistry: theory and practice. Oxford University Press, New York
9. Anastas PT, Zimmerman JB (2003) Design through the 12 principles of green engineering. Environ Sci Technol 37:94–101
10. Dubé MA, Salehpour S (2014) Applying the principles of green chemistry to polymer production technology. Macromol React Eng 8:7–28. doi:10.1002/mren.201300103
11. Odian G (2004) Principles of polymerization. Wiley, Hoboken
12. Williams CK, Hillmyer MA (2008) Polymers from renewable resources: a perspective for a special issue of polymer reviews. Polym Rev 48:1–10. doi:10.1080/15583720701834133
13. Autian J (1975) Structure-toxicity relationships of acrylic monomers. Environ Health Perspect 11:141–152. doi:10.2307/3428337
14. Leggat PA, Kedjarune U (2003) Toxicity of methyl methacrylate in dentistry. Int Dent J 53:126–131. doi:10.1111/j.1875-595X.2003.tb00736.x
15. Zondlo M (2002) Final report on the safety assessment of acrylates copolymer and 33 related cosmetic ingredients. Int J Toxicol 21(Suppl 3):1–50. doi:10.1080/10915810290169800
16. Belgacem MN, Gandini A (2008) Monomers, polymers and composites from renewable resources. Elsevier, Oxford
17. Zhang Y, Dubé MA (2014) Copolymerization of n-butyl methacrylate and d-limonene. Macromol React Eng 8:805–812. doi:10.1002/mren.201400023
18. Zhang C, Yan M, Cochran EW, Kessler MR (2015) Biorenewable polymers based on acrylated epoxidized soybean oil and methacrylated vanillin. Mater Today Commun 5:18–22. doi:10.1016/j.mtcomm.2015.09.003

19. Silvestre AJD, Gandini A (2008) Terpenes: major sources, properties and applications. In: Gandini A (ed) Monomers, Polymers and composites from renewable resources. Elsevier, Amsterdam, pp. 17–38
20. Wilbon PA, Chu F, Tang C (2013) Progress in renewable polymers from natural terpenes, terpenoids, and rosin. Macromol Rapid Commun 34:8–37. doi:10.1002/marc.201200513
21. Roberts W, Day A (1950) A study of the polymerization of alpha-pinene and beta-pinene with friedel crafts type catalysts. J Am Chem Soc 72:1226–1230. doi:10.1021/ja01159a044
22. Zhang Y (2014) Copolymerization of limonene. Dissertation, University of Ottawa, Canada
23. Ramos AM, Lobo LS, Bordado JM (1998) Polymers from pine gum components: radical and coordination homo and copolymerization of pinenes. Macromol Symp 127:43–50. doi:10.1002/masy.19981270109
24. Singh A, Kamal M (2012) Synthesis and characterization of polylimonene: polymer of an optically active terpene. J Appl Polym Sci 125:1456–1459. doi:10.1002/app.36250
25. Lincoln DE, Lawrence BM (1984) The volatile constituents of camphorweed, Heterotheca subaxillaris. Phytochemistry 23:933–934. doi:10.1016/S0031-9422(00)85073-6
26. Corma A, Iborra S, Velty A (2007) Chemical routes for the transformation of biomass into chemicals. Chem Rev 107:2411–2502. doi:10.1021/cr050989d
27. Paz-Pazos M, Pugh C (2006) Synthesis of optically active copolymers of 2,3,4,5,6-pentafluorostyrene and β-pinene with low surface energies. J Polym Sci Part Polym Chem 44:3114–3124. doi:10.1002/pola.21392
28. Wang Y, Li A-L, Liang H, Lu J (2006) Reversible addition–fragmentation chain transfer radical copolymerization of β-pinene and methyl acrylate. Eur Polym J 42:2695–2702. doi:10.1016/j.eurpolymj.2006.06.015
29. Li A-L, Wang Y, Liang H, Lu J (2006) Controlled radical copolymerization of β-pinene and acrylonitrile. J Polym Sci Part Polym Chem 44:2376–2387. doi:10.1002/pola.21316
30. Li A-L, Wang X-Y, Liang H, Lu J (2007) Controlled radical copolymerization of β-pinene and n-butyl acrylate. React Funct Polym 67:481–488. doi:10.1016/j.reactfunctpolym.2007.03.002
31. Yamamoto D, Matsumoto A (2012) Penultimate unit and solvent effects on 2:1 sequence control during radical copolymerization of n-phenylmaleimide with β-pinene. Macromol Chem Phys 213:2479–2485. doi:10.1002/macp.201200421
32. Surburg H, Panten J (2006) Common fragrance and flavor materials. Wiley-VCH, Weinheim, p. 52
33. Mohammad A, Inamuddin (2012) Green solvents. I. Properties and application in chemistry. Springer, New York
34. Maślińska-solich J, Kupka T, Kluczka M, Solich A (1994) Optically active polymers, 2. Copolymerization of limonene with maleic anhydride. Macromol Chem Phys 195:1843–1850. doi:10.1002/macp.1994.021950531
35. Sharma S, Srivastava A (2004) Synthesis and characterization of copolymers of limonene with styrene initiated by azobisisobutyronitrile. Eur Polym J 40:2235–2240. doi:10.1016/j.eurpolymj.2004.02.028
36. Sharma S, Srivastava A (2003) Radical copolymerization of limonene with acrylonitrile: kinetics and mechanism. Polym-Plast Technol Eng 42:485–502. doi:10.1081/PPT-120017966
37. Sharma S, Srivastava AK (2007) Azobisisobutyronitrile-initiated free-radical copolymerization of limonene with vinyl acetate: synthesis and characterization. J Appl Polym Sci 106:2689–2695. doi:10.1002/app.24205
38. Zhang Y, Dubé MA (2014) Copolymerization of 2-ethylhexyl acrylate and d-limonene. Polym Plast Technol Eng 54:499. doi:10.1080/03602559.2014.961080
39. Ren S, Trevino E, Dubé MA (2015) Copolymerization of limonene with n-butyl acrylate. Macromol React Eng 9:339–349. doi:10.1002/mren.201400068

40. Dubé MA, Soares J, Penlidis A, Hamielec AE (1997) Mathematical modeling of multicomponent chain-growth polymerizations in batch, semibatch, and continuous reactors: a review. Ind Eng Chem Res 36:966–1015

41. Bolton JM, Hillmyer MA, Hoye TR (2014) Sustainable thermoplastic elastomers from terpene-derived monomers. ACS Macro Lett 3:717–720. doi:10.1021/mz500339h

42. Horrillo-Martínez P, Virolleaud M-A, Jaekel C (2010) Selective palladium-catalyzed dehydrogenation of limonene to dimethylstyrene. ChemCatChem 2:175–181. doi:10.1002/cctc.200900200

43. United State Department of Agriculture (2016) Oilseeds: World markets and trade. http://www.fas.usda.gov/data/oilseeds-world-markets-and-trade. Accessed 4 Aug 2016

44. Belgacem MN, Gandini A (2008) Materials from vegetable oils: major sources, properties and applications. In: Gandini A (ed) Monomers, polymers and composites from renewable resources. Elsevier, Amsterdam, pp. 39–66

45. Ingram AR, Zupanc AJ, Nicholson HL (1967) Expandable styrene polymers. US Patent 3,359,219, 19 Dec 1967

46. Fernandez AM, Conde A (1983) Monomer reactivity ratios of tung oil and styrene in copolymerization. In: Carraher Jr CE, Sperling LH (eds) Polymer applications of renewable-resource materials. Springer, Berlin, pp 289–302

47. Li FK, Larock RC (2003) Synthesis, structure and properties of new tung oil-styrene-divinylbenzene copolymers prepared by thermal polymerization. Biomacromolecules 4:1018–1025. doi:10.1021/bm034049j

48. Kundu PP, Larock RC (2005) Novel conjugated linseed oil-styrene-divinylbenzene copolymers prepared by thermal polymerization. 1. Effect of monomer concentration on the structure and properties. Biomacromolecules 6:797–806. doi:10.1021/bm049429z

49. Henna PH, Andjelkovic DD, Kundu PP, Larock RC (2007) Biobased thermosets from the free-radical copolymerization of conjugated linseed oil. J Appl Polym Sci 104:979–985. doi:10.1002/app.25788

50. Gultekin M, Beker U, Güner FS, et al (2000) Styrenation of castor oil and linseed oil by macromer method. Macromol Mater Eng 283:15–20. doi:10.1002/1439-2054(20001101)283:1<15::AID-MAME15>3.0.CO;2-I

51. Akbas T, Beker ÜG, Güner FS, et al (2003) Drying and semidrying oil macromonomers. III. Styrenation of sunflower and linseed oils. J Appl Polym Sci 88:2373–2376. doi:10.1002/app.11638

52. Eren T, Küsefoğlu SH (2004) Synthesis and polymerization of the bromoacrylated plant oil triglycerides to rigid, flame-retardant polymers. J Appl Polym Sci 91:2700–2710. doi:10.1002/app.13471

53. Hernandez S, Vigueras E (2013) Acrylated-epoxidized soybean oil-based polymers and their use in the generation of electrically conductive polymer composites. In: El-Shemy H (ed) Soybean – bio-active compounds. InTech, pp 231–263

54. Pelletier H, Belgacem N, Gandini A (2006) Acrylated vegetable oils as photocrosslinkable materials. J Appl Polym Sci 99:3218–3221. doi:10.1002/app.22322

55. Ahn BK, Sung J, Rahmani N, et al (2013) UV-curable, high-shear pressure-pensitive adhesives derived from acrylated epoxidized soybean oil. J Adhes 89:323–338. doi:10.1080/00218464.2013.749102

56. Can E, Wool RP, Küsefoğlu S (2006) Soybean- and castor-oil-based thermosetting polymers: mechanical properties. J Appl Polym Sci 102:1497–1504. doi:10.1002/app.24423

57. Can E, Wool RP, Küsefoğlu S (2006) Soybean and castor oil based monomers: synthesis and copolymerization with styrene. J Appl Polym Sci 102:2433–2447. doi:10.1002/app.24548

58. Roberge S, Dubé MA (2016) Bulk terpolymerization of conjugated linoleic acid with styrene and butyl acrylate. ACS Sustain Chem Eng 4:264–272. doi:10.1021/acssuschemeng.5b01106

59. Roberge S, Dubé MA (2016) Emulsion-based pressure sensitive adhesives from conjugated linoleic acid/styrene/butyl acrylate terpolymers. Int J Adhes Adhes 70:17–25. doi:10.1016/j.ijadhadh.2016.05.003

60. Vendamme R, Schüwer N, Eevers W (2014) Recent synthetic approaches and emerging bio-inspired strategies for the development of sustainable pressure-sensitive adhesives derived from renewable building blocks. J Appl Polym Sci. doi:10.1002/app.40669

61. Anderson KS, Lewandowski KM, Fansler DD et al (2011) 2-octyl (meth)acrylate adhesive composition. US Patent 7,893,179, 22 Feb 2011

62. Çayli G, Meier MAR (2008) Polymers from renewable resources: bulk ATRP of fatty alcohol-derived methacrylates. Eur J Lipid Sci Technol 110:853–859. doi:10.1002/ejlt. 200800028

63. Wool RP, Bunker SP (2003) Pressure sensitive adhesives from plant oils. US Patent 6,646,033, 11 Nov 2003

64. Christoph R, Schmidt B, Steinberner U, et al (2000) Glycerol. In: Ullmann's encyclopedia of industrial chemistry. Wiley-VCH Verlag GmbH & Co. KGaA

65. Salehpour S, Dubé MA (2011) Towards the sustainable production of higher-molecular-weight polyglycerol. Macromol Chem Phys 212:1284–1293. doi:10.1002/macp.201100064

66. Liu Y, Tüysüz H, Jia C-J, et al (2010) From glycerol to allyl alcohol: iron oxide catalyzed dehydration and consecutive hydrogen transfer. Chem Commun 46:1238–1240. doi:10.1039/B921648K

67. Zhang H, Grinstaff MW (2014) Recent advances in glycerol polymers: chemistry and biomedical applications. Macromol Rapid Commun 35:1906–1924. doi:10.1002/marc. 201400389

68. Pham PD, Monge S, Lapinte V, et al (2013) Various radical polymerizations of glycerol-based monomers. Eur J Lipid Sci Technol 115:28–40. doi:10.1002/ejlt.201200202

69. Roice M, Pillai VNR (2005) Poly(styrene-co-glycerol dimethacrylate): synthesis, characterization, and application as a resin for gel-phase peptide synthesis. J Polym Sci Part Polym Chem 43:4382–4392. doi:10.1002/pola.20917

70. Vijitha K, Dhanya K, Francis B, et al (2009) Synthesis and characterization of glycerol dimethacrylate-4-vinyl pyrrole. Asian J Chem 21:6811–6818

71. Miranda LN, Ford WT (2005) Binary copolymer reactivity of tert-butyl methacrylate, 2-(N, N-dimethylamino)ethyl methacrylate, solketal methacrylate, and 2-bromoethyl methacrylate. J Polym Sci Part Polym Chem 43:4666–4669. doi:10.1002/pola.20939

72. Iio K, Kobayashi K, Matsunaga M (2007) Radical polymerization of allyl alcohol and allyl acetate. Polym Adv Technol 18:953–958. doi:10.1002/pat.870

73. Galbis JA, de Gracia G-MM, Violante de Paz M, Galbis E (2016) Synthetic polymers from sugar-based monomers. Chem Rev 116:1600–1636. doi:10.1021/acs.chemrev.5b00242

74. Galbis JA, García-Martín MG (2008) Sugars as monomers. In: Gandini A, Belgacem M (eds) Monomers, polymers and composites from renewable resources. Elsevier, Amsterdam, pp 89–114

75. Lavilla C, Alla A, de Ilarduya AM, et al (2012) Carbohydrate-based copolyesters made from bicyclic acetalized galactaric acid. J Polym Sci Part Polym Chem 50:1591–1604. doi:10. 1002/pola.25930

76. Lavilla C, Alla A, de Ilarduya AM, et al (2012) Bio-based poly(butylene terephthalate) copolyesters containing bicyclic diacetalized galactitol and galactaric acid: influence of composition on properties. Polymer 53:3432–3445. doi:10.1016/j.polymer.2012.05.048

77. Wu J, Eduard P, Thiyagarajan S, et al (2012) Semicrystalline polyesters based on a novel renewable building block. Macromolecules 45:5069–5080. doi:10.1021/ma300782h

78. Engler AC, Ke X, Gao S, et al (2015) Hydrophilic polycarbonates: promising degradable alternatives to poly(ethylene glycol)-based stealth materials. Macromolecules 48:1673–1678. doi:10.1021/acs.macromol.5b00156

79. Feng J, Zhuo R-X, Zhang X-Z (2012) Construction of functional aliphatic polycarbonates for biomedical applications. Prog Polym Sci 37:211–236. doi:10.1016/j.progpolymsci.2011.07. 008

80. Begines B, Zamora F, de Paz MV, et al (2015) Polyurethanes derived from carbohydrates and cystine-based monomers. J Appl Polym Sci. doi:10.1002/app.41304

81. Boyer A, Lingome CE, Condassamy O, et al (2013) Glycolipids as a source of polyols for the design of original linear and cross-linked polyurethanes. Polym Chem 4:296–306. doi:10.1039/c2py20588b

82. Reyes-Mercado Y, Vázquez F, Rodríguez-Gómez FJ, Duda Y (2008) Effect of the acrylic acid content on the permeability and water uptake of poly(styrene-co-butyl acrylate) latex films. Colloid Polym Sci 286:603–609. doi:10.1007/s00396-008-1838-6

83. Datta R, Henry M (2006) Lactic acid: recent advances in products, processes and technologies – a review. J Chem Technol Biotechnol 81:1119–1129. doi:10.1002/jctb.1486

84. Xu X, Lin J, Cen P (2006) Advances in the research and development of acrylic acid production from biomass. Chin J Chem Eng 14:419–427. doi:10.1016/S1004-9541(06)60094-3

85. de Guzman D (2012) Bio-acrylic acid on the way. In: Green Chem. Blog. http://greenchemicalsblog.com/2012/09/01/5060/. Accessed 17 Aug 2016

86. Dishisha T, Pyo S-H, Hatti-Kaul R (2015) Bio-based 3-hydroxypropionic- and acrylic acid production from biodiesel glycerol via integrated microbial and chemical catalysis. Microb Cell Factories 14:200. doi:10.1186/s12934-015-0388-0

87. Burk MJ, Pharkya P, Dien SJV et al (2012) Methods for the synthesis of olefins and derivatives. US Patent 20,120,094,341, 19 Apr 2012

88. Green EM (2011) Fermentative production of butanol – the industrial perspective. Curr Opin Biotechnol 22:337–343. doi:10.1016/j.copbio.2011.02.004

89. de Guzman D (2013) Bio-MMA development expands. In: Green Chem. Blog. http://greenchemicalsblog.com/2013/03/14/bio-mma-development-expands/. Accessed 17 Aug 2016

90. University of Minnesota (2011) Biological pathways produce isobutyric acid using renewable resources. http://license.umn.edu/technologies/20110077_biological-pathways-produce-isobutyric-acid-using-renewable-resources. Accessed 22 Aug 2016

91. Bloom PD, Venkitasubramanian P (2009) Monomers and polymers from bioderived carbon. US Patent 20,090,018,300, 8 Jul 2008

92. Bloembergen S, McLennan IJ, Narayan R (1999) Sugar based vinyl monomers and copolymers useful in repulpable adhesives and other applications. US Patent 5,872,199, 16 Feb 1999

93. Bloembergen S, McLennan IJ, Narayan R (2001) Environmentally friendly sugar-based vinyl monomers useful in repulpable adhesives and other applications. US Patent 6,242,593, 5 Jun 2001

94. Dunn AS (1986) Polymeric stabilization of colloidal dispersions. Polym Int J 18:278–278. doi:10.1002/pi.4980180420

95. Thickett SC, Gilbert RG (2007) Emulsion polymerization: state of the art in kinetics and mechanisms. Polymer 48:6965–6991. doi:10.1016/j.polymer.2007.09.031

96. Zecha H (1981) Stabilization of colloidal dispersions by polymer adsorption. Acta Polym 32:582–582. doi:10.1002/actp.1981.010320915

97. Kronberg B, Holmberg K, Lindman B (2014) Environmental and health aspects of surfactants. In: Surface chemistry of surfactants and polymers. Wiley, Hoboken, pp. 49–64

98. Liwarska-Bizukojc E, Miksch K, Malachowska-Jutsz A, Kalka J (2005) Acute toxicity and genotoxicity of five selected anionic and nonionic surfactants. Chemosphere 58:1249–1253. doi:10.1016/j.chemosphere.2004.10.031

99. von Rybinski W, Hill K (1998) Alkyl polyglycosides – properties and applications of a new class of surfactants. Angew Chem Int Ed 37:1328–1345. doi:10.1002/(SICI)1521-3773(19980605)37:103.0.CO;2-9

100. Holmberg K (2003) Novel surfactants: preparation, applications, and biodegradability, 2nd edn. Marcel Dekker, New York

101. Benvegnu T, Plusquellec D, Lemiègre L (2008) Surfactants from renewable sources: synthesis and applications. In: Gandini A (ed) Monomers, polymers and composites from renewable resources. Elsevier, Amsterdam, pp. 153–178

102. Hill K (2010) Surfactants based on carbohydrates and proteins for consumer products and technical applications. In: Kjellin M, Johansson I (eds) Surfactants from renewable resources. Wiley, Hoboken, pp. 63–84

103. Global Market Insights (2016) Alkyl polyglucosides (APG) biosurfactants biosurfactants market size https://www.gminsights.com/industry-analysis/alkyl-polyglucosides-apg-biosurfactants-market. Accessed 7 Sept 2016

104. Aulmann W, Sterzel W (1996) Toxicology of alkyl polyglycosides. In: Hill K, von Rybinski W, Stoll G (eds) Alkyl polyglycosides. Wiley-VCH, Weinheim, pp. 151–167

105. Lazaridis N, Alexopoulos AH, Kiparissides C (2001) Semi-batch emulsion copolymerization of vinyl acetate and butyl acrylate using oligomeric nonionic surfactants. Macromol Chem Phys 202:2614–2622. doi:10.1002/1521-3935(20010801)202:123.0.CO;2-E

106. Chen L (2012) Application of green commercial surfactant in preparing purely acrylic latex via semi-continuous seeded emulsion polymerization. J Surfactant Deterg 16:197–202. doi:10.1007/s11743-012-1373-9

107. Klima R, Pippin WH, Natale M et al (2000) Alkylpolyglycoside containing surfactant blends for emulsion polymerization. US Patent 6,117,934, 12 Sep 2000

108. Maver TL, Krasnansky R (2001) Aqueous coating composition with improved block resistance containing alkyl polyglycoside surfactant mixtures. US Patent 6,117,934, 12 Sep 2000

109. Hoydonckx HE, Vos DED, Chavan SA, Jacobs PA (2004) Esterification and transesterification of renewable chemicals. Top Catal 27:83–96. doi:10.1023/B:TOCA.0000013543.96438.1a

110. Stockburger GJ (1981) Process for preparing sorbitan esters. US Patent 4,297,290, 27 Oct 1981

111. Ellis JMH, Lewis JJ, Beattie RJ (1998) Manufacture of fatty acid esters of sorbitan as surfactants. WO Patent 1,998,004,540, 5 Feb 1998

112. Milstein N (1992) Improved esterification of oxyhydrocarbon polyols and ethers thereof, and products therefrom. WO Patent 1,992,000,947, 23 Jan 1992

113. Falbe J (ed) (1987) Surfactants in consumer products. Springer, Berlin

114. Kovačič S, Matsko NB, Jerabek K, et al (2012) On the mechanical properties of HIPE templated macroporous poly(dicyclopentadiene) prepared with low surfactant amounts. J Mater Chem A 1:487–490. doi:10.1039/C2TA00546H

115. Silverstein MS (2014) Emulsion-templated porous polymers: a retrospective perspective. Polymer 55:304–320. doi:10.1016/j.polymer.2013.08.068

116. Capek I, Chudej J (1999) On the fine emulsion polymerization of styrene with non-ionic emulsifier. Polym Bull 43:417–424. doi:10.1007/s002890050630

117. Giovannoli C, Passini C, Anfossi L, et al (2015) Comparison of binding behavior for molecularly imprinted polymers prepared by hierarchical imprinting or Pickering emulsion polymerization. J Sep Sci 38:3661–3668. doi:10.1002/jssc.201500511

118. Yao F, Yan G-C, Xu L-Q, et al (2014) Hairy fluorescent nanoparticles from one-pot click chemistry and atom transfer radical emulsion polymerization. Polym Int 63:237–243. doi:10.1002/pi.4491

119. Clark E (1988) Inverse emulsion polymerization with sorbitan fatty acid esters and ethoxylated alcohol. US Patent 4,764,574, 16 Aug 1988

120. Ramli RA, Hashim S, Laftah WA (2013) Synthesis, characterization, and morphology study of poly(acrylamide-co-acrylic acid)-grafted-poly(styrene-co-methyl methacrylate) "raspberry"-shape like structure microgels by pre-emulsified semi-batch emulsion polymerization. J Colloid Interface Sci 391:86–94. doi:10.1016/j.jcis.2012.09.047

121. Wan T, Zang T, Wang Y, et al (2010) Preparation of water soluble Am–AA–SSS copolymers by inverse microemulsion polymerization. Polym Bull 65:565–576. doi:10.1007/s00289-009-0234-9

122. Zhang Y, Li T, Jin Z, et al (2007) Synthesis of nanoiron by microemulsion with span/tween as mixed surfactants for reduction of nitrate in water. Front Environ Sci Eng China 1:466–470. doi:10.1007/s11783-007-0074-5

123. Mollet H, Grubenmann A (2000) Emulsions – properties and production. In: Formulation technology. Wiley-VCH, Weinheim, pp 59–104
124. Osipow L, Snell FD, York WC, Finchler A (1956) Methods of preparation fatty acid esters of sucrose. Ind Eng Chem 48:1459–1462. doi:10.1021/ie51400a026
125. Parker KJ, Khan RA, Mufti KS (1976) Process of making sucrose esters. US Patent 3,996,206, 7 Dec 1976
126. Reuben F, Theodore W, Hampden Z (1973) Process for the production of sucrose esters of fatty acids. US Patent 3,714,144, 30 Jan1973
127. Markets and Markets (2016) Sucrose esters market by application, form, region – 2020. http://www.marketsandmarkets.com/Market-Reports/sucrose-esters-market-191170937. html. Accessed 12 Sept 2016
128. Crandall MD, Nelson RL (1995) Nonionic, pH-neutral pressure sensitive adhesive US Patent 4,424,122, 13 Jun 1995
129. Nagasuna K, Namba T, Miyake K et al (1990) Production process for water-absorbent resin. US Patent 4973,362, 27 Nov 1990
130. Lan Z, Daga R, Whitehouse R, et al (2014) Structure–properties relations in flexible polyurethane foams containing a novel bio-based crosslinker. Polymer 55:2635–2644. doi:10.1016/j.polymer.2014.03.061
131. Ma S, Jiang Y, Liu X, et al (2014) Bio-based tetrafunctional crosslink agent from gallic acid and its enhanced soybean oil-based UV-cured coatings with high performance. RSC Adv 4:23036. doi:10.1039/c4ra01311e
132. Oprea S (2009) Synthesis and properties of polyurethane elastomers with castor oil as crosslinker. J Am Oil Chem Soc 87:313–320. doi:10.1007/s11746-009-1501-5
133. Ding C, Shuttleworth PS, Makin S, et al (2015) New insights into the curing of epoxidized linseed oil with dicarboxylic acids. Green Chem 17:4000–4008. doi:10.1039/C5GC00912J
134. Supanchaiyamat N, Shuttleworth PS, Hunt AJ, et al (2012) Thermosetting resin based on epoxidised linseed oil and bio-derived crosslinker. Green Chem 14:1759–1765. doi:10.1039/C2GC35154D
135. Mathers RT, Damodaran K (2007) Renewable chain transfer agents for metallocene polymerizations: the effects of chiral monoterpenes on the polyolefin molecular weight and isotacticity. J Polym Sci Part Polym Chem 45:3150–3165. doi:10.1002/pola.22111
136. Mathers RT, McMahon KC, Damodaran K, et al (2006) Ring-opening metathesis polymerizations in d-limonene: a renewable polymerization solvent and chain transfer agent for the synthesis of alkene macromonomers. Macromolecules 39:8982–8986. doi:10.1021/ma061699h
137. Clark JH (1995) Chemistry of waste minimization. Blackie Academic & Professional, New York
138. Nomura Y, Teshima W, Kawahara T, et al (2006) Genotoxicity of dental resin polymerization initiators in vitro. J Mater Sci Mater Med 17:29–32. doi:10.1007/s10856-006-6326-2
139. Caillol S (2014) Lifecycle assessment and green chemistry: a look at innovative tools for sustainable development. In: Hamaide T, Deterre R, Feller J-F (eds) Environmental impact of polymers. Wiley, Hoboken, pp. 65–89
140. Araújo PHH, Sayer C, Giudici R, Poço JGR (2002) Techniques for reducing residual monomer content in polymers: a review. Polym Eng Sci 42:1442–1468. doi:10.1002/pen.11043
141. Oka T, Tsubota K, Shinjo T et al (1999) Process for preparing solvent-type acrylic pressure-sensitive adhesives and medical pressure-sensitive adhesive. US Patent 5,886,122, 26 May1996
142. Heider L, Storck G, Weintz H-J (1992) Preparation of polymers from olefinically unsaturated monomers. US Patent 5,087,676, 11 Feb 1992
143. Minematsu H, Matsumoto K, Saeki T, Kishi A (1981) Low residual monomer α-methylstyrene-acrylonitrile copolymers and ABS blends therefrom. US Patent 4,294,946, 13 Oct 1981

144. Humme G, Plato H, Ott K-H et al (1983) Process for the removal of residual monomers from ABS polymers. US Patent 4,399,273, 16 Aug 1983
145. Aerts M, Meuldijk J, Kemmere M, Keurentjes J (2011) Residual monomer reduction in polymer latex products by extraction with supercritical carbon dioxide. Macromol Symp 302:297–304. doi:10.1002/masy.201000052
146. Schull V, Arnoldi D (1998) Method of producing non-vitrified processing aid low in residual monomers for thermoplastic polymers. US Patent 5,767,231, 16 Jun 1998
147. Heinze C, Botsch F, Wolff H (1981) Process and device for continuously treating with gases aqueous dispersions of polyvinyl chloride. US Patent 4,301,275, 17 Nov 1981
148. Copelli S, Derudi M, Sempere J, et al (2011) Emulsion polymerization of vinyl acetate: safe optimization of a hazardous complex process. J Hazard Mater 192:8–17. doi:10.1016/j.jhazmat.2011.04.066
149. Fonseca GE, Dubé MA, Penlidis A (2009) A critical overview of sensors for monitoring polymerizations. Macromol React Eng 3:327–373. doi:10.1002/mren.200900024
150. Chen M, Reichert K-H (1993) Studies on free radical polymerization by adiabatic reaction calorimetry. Polym React Eng 1:145–170. doi:10.1080/10543414.1992.10744426
151. Goikoetxea M, Heijungs R, Barandiaran MJ, Asua JM (2008) Energy efficient emulsion polymerization strategies. Macromol React Eng 2:90–98. doi:10.1002/mren.200700042
152. Wang S, Daniels ES, Sudol ED, et al (2016) Isothermal emulsion polymerization of n-butyl methacrylate with KPS and redox initiators: kinetic study at different surfactant/initiator concentrations and reaction temperature. J Appl Polym Sci 133:43037. doi:10.1002/app.43037
153. Garg DK, Serra CA, Hoarau Y, et al (2014) Analytical solution of free radical polymerization: applications-implementing nonisothermal effect. Macromolecules 47:8514–8523. doi:10.1021/ma501964h
154. Pohl K, Rodriguez F (1981) Adiabatic polymerization of acrylamide using a persulfate–bisulfite redox couple. J Appl Polym Sci 26:611–618. doi:10.1002/app.1981.070260220
155. Thomson RAM (1986) A kinetic study of the adiabatic polymerization of acrylamide. J Chem Educ 63:362. doi:10.1021/ed063p362
156. Tonoyan AO, Leikin AD, Davtyan SP, et al (1973) Kinetics of the adiabatic polymerization of methyl methacrylate. Polym Sci USSR 15:2080–2085. doi:10.1016/0032-3950(73)90424-3
157. Wang S (2013) Redox-initiated adiabatic emulsion polymerization. Dissertation, Lehigh University
158. Chemtob A, Kunstler B, Croutxé-Barghorn C, Fouchard S (2010) Photoinduced miniemulsion polymerization. Colloid Polym Sci 288:579–587. doi:10.1007/s00396-010-2190-1
159. Mah S, Koo D, Jeon H, Kwon S (2002) Photo-induced emulsion polymerization of vinyl acetate in the presence of poly(oxyethylene)10 nonyl phenyl ether ammonium sulfate, an anionic emulsifier (I). J Appl Polym Sci 84:2425–2431. doi:10.1002/app.10531
160. Turro NJ, Chow M-F, Chung C-J, Tung C-H (1980) An efficient, high conversion photoinduced emulsion polymerization. Magnetic field effects on polymerization efficiency and polymer molecular weight. J Am Chem Soc 102:7391–7393. doi:10.1021/ja00544a053
161. Aldana-García MA, Palacios J, Vivaldo-Lima E (2005) Modeling of the microwave initiated emulsion polymerization of styrene. J Macromol Sci A 42:1207–1225. doi:10.1080/10601320500189505
162. Ergan BT, Bayramoğlu M, Özcan S (2015) Emulsion polymerization of styrene under continuous microwave irradiation. Eur Polym J 69:374–384. doi:10.1016/j.eurpolymj.2015.06.021
163. Zhu X, Chen J, Cheng Z, et al (2003) Emulsion polymerization of styrene under pulsed microwave irradiation. J Appl Polym Sci 89:28–35. doi:10.1002/app.12089

164. Bhanvase BA, Pinjari DV, Sonawane SH, et al (2012) Analysis of semibatch emulsion polymerization: role of ultrasound and initiator. Ultrason Sonochem 19:97–103. doi:10. 1016/j.ultsonch.2011.05.016
165. Cheung HM, Gaddam K (2000) Ultrasound-assisted emulsion polymerization of methyl methacrylate and styrene. J Appl Polym Sci 76:101–104. doi:10.1002/(SICI)1097-4628 (20000404)76:1<101::AID-APP13>3.0.CO;2-F
166. Chou HCJ, Stoffer JO (1999) Ultrasonically initiated free radical-catalyzed emulsion polymerization of methyl methacrylate (i). J Appl Polym Sci 72:797–825. doi:10.1002/(SICI) 1097-4628(19990509)72:6<797::AID-APP7>3.0.CO;2-Z
167. Korkut I, Bayramoglu M (2014) Various aspects of ultrasound assisted emulsion polymerization process. Ultrason Sonochem 21:1592–1599. doi:10.1016/j.ultsonch.2013.12.028
168. Xia H, Wang Q, Qiu G (2003) Polymer-encapsulated carbon nanotubes prepared through ultrasonically initiated in situ emulsion polymerization. Chem Mater 15:3879–3886. doi:10. 1021/cm0341890
169. Chien DCH, Penlidis A (1990) On-line sensors for polymerization reactors. J Macromol Sci Part C 30:1–42. doi:10.1080/07366579008050904
170. Elizalde O, Azpeitia M, Reis MM, et al (2005) Monitoring emulsion polymerization reactors: calorimetry versus Raman spectroscopy. Ind Eng Chem Res 44:7200–7207. doi:10.1021/ ie050451y
171. Gesthuisen R, Krämer S, Niggemann G, et al (2005) Determining the best reaction calorimetry technique: theoretical development. Comput Chem Eng 29:349–365. doi:10.1016/j. compchemeng.2004.10.009
172. Frauendorfer E, Wolf A, Hergeth W-D (2010) Polymerization online monitoring. Chem Eng Technol 33:1767–1778. doi:10.1002/ceat.201000265
173. Liu J, Tai H, Howdle SM (2005) Precipitation polymerisation of vinylidene fluoride in supercritical CO2 and real-time calorimetric monitoring. Polymer 46:1467–1472. doi:10. 1016/j.polymer.2004.12.015
174. Wang W, Griffiths RMT, Giles MR, et al (2003) Monitoring dispersion polymerisations of methyl methacrylate in supercritical carbon dioxide. Eur Polym J 39:423–428. doi:10.1016/ S0014-3057(02)00249-5
175. De Buruaga IS, Echevarría A, Armitage PD, et al (1997) On-line control of a semibatch emulsion polymerization reactor based on calorimetry. AICHE J 43:1069–1081. doi:10.1002/ aic.690430420
176. Lamb DJ, Fellows CM, Morrison BR, Gilbert RG (2005) A critical evaluation of reaction calorimetry for the study of emulsion polymerization systems: thermodynamic and kinetic aspects. Polymer 46:285–294. doi:10.1016/j.polymer.2004.11.026
177. Rincón FD, Esposito M, de Araújo PHH, et al (2013) Calorimetric estimation employing the unscented kalman filter for a batch emulsion polymerization reactor. Macromol React Eng 7:24–35. doi:10.1002/mren.201200044
178. Rincón FD, Esposito M, de Araújo PHH, et al (2014) Robust calorimetric estimation of semi-continuous and batch emulsion polymerization systems with covariance estimation. Macromol React Eng 8:456–466. doi:10.1002/mren.201300151
179. De La Rosa LV, Sudol ED, El-Aasser MS, Klein A (1999) Emulsion polymerization of styrene using reaction calorimeter.II. Importance of maximum in rate of polymerization. J Polym Sci Part Polym Chem 37:4066–4072
180. Vicente M, BenAmor S, Gugliotta LM, et al (2001) Control of molecular weight distribution in emulsion polymerization using on-line reaction calorimetry. Ind Eng Chem Res 40:218–227. doi:10.1021/ie000387e
181. Blythe PJ, Klein A, Sudol ED, El-Aasser MS (1999) Enhanced droplet nucleation in styrene miniemulsion polymerization. 3. Effect of shear in miniemulsions that use cetyl alcohol as the cosurfactant. Macromolecules 32:4225–4231. doi:10.1021/ma981977f
182. Blythe PJ, Klein A, Sudol ED, El-Aasser MS (1999) Enhanced droplet nucleation in styrene miniemulsion polymerization. 2. Polymerization kinetics of homogenized emulsions

containing predissolved polystyrene. Macromolecules 32:6952–6957. doi:10.1021/ma981976n

183. Blythe PJ, Morrison BR, Mathauer KA, et al (1999) Enhanced droplet nucleation in styrene miniemulsion polymerization. 1. Effect of polymer type in sodium lauryl sulfate/cetyl alcohol miniemulsions. Macromolecules 32:6944–6951. doi:10.1021/ma981975v

184. Vieira RAM, Sayer C, Lima EL, Pinto JC (2002) In-line and in situ monitoring of semi-batch emulsion copolymerizations using near-infrared spectroscopy. J Appl Polym Sci 84:2670–2682. doi:10.1002/app.10434

185. Dubé MA, Li L (2010) In-line monitoring of SBR emulsion polymerization using ATR-FTIR spectroscopy. Polym Plast Technol Eng 49:648–656. doi:10.1080/03602551003664909

186. Hua H, Dubé MA (2002) In-line monitoring of emulsion homo- and copolymerizations using ATR-FTIR spectrometry. Polym React Eng 10:21–39. doi:10.1081/PRE-120002903

187. Poljanšek I, Fabjan E, Burja K, Kukanja D (2013) Emulsion copolymerization of vinyl acetate-ethylene in high pressure reactor-characterization by inline FTIR spectroscopy. Prog Org Coat 76:1798–1804. doi:10.1016/j.porgcoat.2013.05.019

188. Roberge S, Dubé MA (2016) Infrared process monitoring of conjugated linoleic acid/styrene/butyl acrylate bulk and emulsion terpolymerization. J Appl Polym Sci 133:n/a–n/a. doi:10.1002/app.43574

189. Ampelli C, Di Bella D, Maschio G, Russo A (2006) Calorimetric study of the inhibition of runaway reactions during methylmethacrylate polymerization processes. J Loss Prev Process Ind 19:419–424. doi:10.1016/j.jlp.2005.10.003

190. Morgan AB, Gilman JW (2013) An overview of flame retardancy of polymeric materials: application, technology, and future directions. Fire Mater 37:259–279. doi:10.1002/fam.2128

Adv Polym Sci (2018) 280: 101–120
DOI: 10.1007/12_2017_15
© Springer International Publishing AG 2017
Published online: 25 June 2017

# Development of Novel Materials from Polymerization of Pickering Emulsion Templates

He Zhu, Lei Lei, Bo-Geng Li, and Shiping Zhu

**Abstract** Pickering emulsions are an attractive type of emulsion. These particle-stabilized emulsions possess many advantages that conventional emulsions do not have. Recently, novel hybrid materials derived from Pickering emulsions have become an emerging research topic. These novel structures include spheres, hollow capsules, and porous foams, depending on the design of the Pickering emulsion template. Polymerization is always involved in order to support the structure after removal of the Pickering emulsion template. We present an overview of recent advances in the development of polymeric materials from Pickering emulsion templates. Developments are organized according to the physical morphologies (spheres, capsules, and foams) of materials derived from Pickering emulsions and the particles (inorganic and organic stabilizers) employed in stabilizing the emulsions.

**Keywords** Capsules • Foams • Inorganic particles • Organic particles • Pickering emulsion templates • Polymerization • Spheres

H. Zhu
Department of Chemical Engineering, McMaster University, Hamilton, ON, Canada, L8S 4L7

State Key Lab of Chemical Engineering, College of Chemical and Biological Engineering, Zhejiang University, Hangzhou 310027, China

L. Lei and S. Zhu (✉)
Department of Chemical Engineering, McMaster University, Hamilton, ON, Canada, L8S 4L7
e-mail: shipingzhu@mcmaster.ca

B.-G. Li (✉)
State Key Lab of Chemical Engineering, College of Chemical and Biological Engineering, Zhejiang University, Hangzhou 310027, China
e-mail: bgli@zju.edu.cn

## Contents

# 1  Introduction

Pickering emulsions, in contrast to conventional surfactant-stabilized emulsions, are stabilized by solid particles and were first described by Ramsden [1] and Pickering [2]. With the development of material science, numerous kinds of functional particles have been designed and prepared in recent decades, which greatly increases the number of potential candidates for this type of emulsion. Compared with conventional emulsions, Pickering emulsions provide several advantages. First, the irreversible absorption of particles at the interface leads to highly stable emulsions [3, 4]. Second, the absence of surfactants makes these emulsions more attractive in cosmetic and pharmaceutical applications because some surfactants are considered toxic [5]. Third, the functional particles can bring novel properties that conventional emulsions do not possess.

During preparation of Pickering emulsions, the shear force of emulsification promotes the migration of particulate stabilizers to the liquid–liquid interface to minimize the surface energy of the system and stabilize the dispersed phase droplets against coalescence [3, 6]. Theoretically, any particles that are not completely wetted by liquid phase can be used as stabilizers for Pickering emulsions. In practice, particles that favor water wetting tend to reduce the oil contact area and form oil-in-water (O/W) emulsions. In contrast, lipophilic particles stabilize water-in-oil (W/O) emulsions [3, 7]. The morphology and stability of Pickering emulsions are influenced by many factors in addition to the wettability of particulate stabilizers. Particle size has a profound effect on emulsion stability [3] and particle concentration affects the size of emulsion droplets [8]. Theoretical studies on Pickering emulsions have been described by Binks and Horozov [3, 9].

Both organic and inorganic nanoparticles have been used in preparation of Pickering emulsions. Inorganic particles, including silica, titania, iron oxide, and metal–organic frameworks (MOFs), have been employed as Pickering stabilizers. Of these, silica is the most popular candidate, probably because of its easy functionalization and low cost. Compared with rigid inorganic particles, organic particles (especially soft particles such as microgels, proteins, and bacteria) are more flexible for Pickering emulsion preparation as a result of their interfacial activity, deformability, oil/water compatibility, tailored functionality, etc. [8, 10, 11].

Many papers and books have comprehensively reviewed the history, chemistry, physics, and potential applications of Pickering emulsions [3, 4, 7, 8, 12, 13].

In addition to the preparation of Pickering emulsions, materials derived from Pickering emulsions have become an emerging research topic in recent years. Pickering emulsions provide excellent templates for fabrication of materials with different morphologies, such as solid/porous spheres, hollow capsules, and foams, leading to novel and enhanced properties of the obtained materials. Polymerization is usually involved in the process because it helps maintain the structures after removal of emulsion templates. The particles stay at the oil–water interface in emulsions and remain at the interface once the emulsion template is removed after polymerization, thus functionalizing the surface of the resulting materials and endowing novel properties to the obtained composites. This review summarizes recent progress in the development of materials derived from the polymerization of Pickering emulsions and the possible applications of these materials. There exist a variety of Pickering emulsions so we have organized the contents of this review according to the physical morphology (sphere, capsule, and foam) of material derived from Pickering emulsions and the type of particle (inorganic and organic) employed in stabilizing the emulsion.

## 2 Materials Derived from Pickering Emulsions

Colloid particles self-assemble at the interface of two immiscible liquids, typically water and oil, to form Pickering emulsions. With Pickering emulsions as templates, a variety of materials can be derived by polymerization of part of the emulsion: (1) Polymerizing the dispersed phase droplets generates solid/porous spheres with surfaces coated by colloid particles. (2) Binding (crosslinking, bridging) the colloid particles at the oil–water interface converts Pickering emulsions into robust microcapsules. (3) Polymerizing the continuous phase results in porous matrixes (e.g., foams) with the surface of inner cells (either close or open cells) modified by colloid particles.

### 2.1 Solid/Porous Spheres

Early efforts focused on the fabrication of materials with a polymer core and particle shell. Pickering emulsion templates with either W/O or O/W morphology are used as precursors. Monomers (e.g., styrene [14], methyl acrylate [15], *n*-butyl acrylate [16]) dispersed in (or used as) the core phase are stabilized by organic or inorganic particle stabilizers and then polymerized to generate hybrid latexes. Binding between stabilizer particles and polymer core is usually achieved by introducing different interactions, such as acid–base interactions and electrostatic

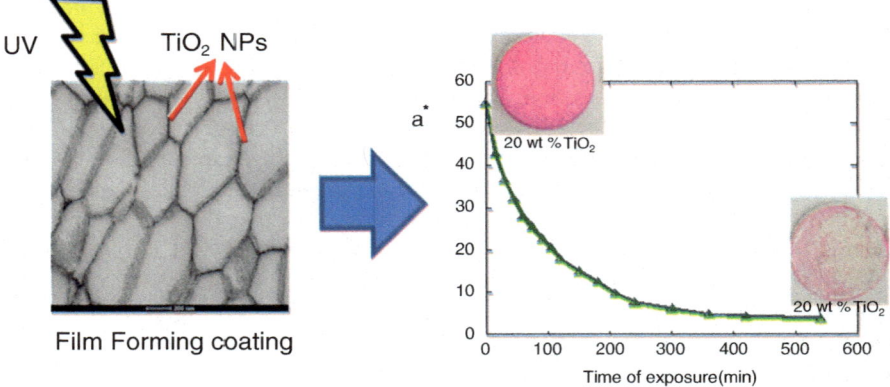

**Fig. 1** TiO₂ films with honeycomb structure and their photocatalytic behavior. Reprinted with permission from [22]

interactions. A recent comprehensive review by Ziener and colleagues summarizes the preparation of Pickering-type latex particles [17].

### 2.1.1 Inorganic Stabilizer

Hybrid latexes have mostly been applied as film-forming composite particles. Armes and coworkers pioneered the preparation of silica-based Pickering emulsions and prepared highly transparent films using film-forming colloidal silica-polymer latexes [18, 19]. Laponite, a type of clay particle, was also used to prepare a latex of armored polymer-Laponite composite particles through Pickering emulsion [20, 21]. The transparent films prepared from careful drying of the latex particles possessed a honeycomb morphology and showed excellent mechanical and thermal properties [20]. Recently, Asua and coworkers reported the fabrication of TiO₂-containing composite films prepared from TiO₂ Pickering-stabilized methyl methacrylate/$n$-butyl acrylate copolymer latexes (Fig. 1) [22]. The films showed excellent self-cleaning activity as a result of the photocatalytic activity of TiO₂.

### 2.1.2 Organic Stabilizer

Microfluidic emulsification is an emerging technology for preparation of highly monodispersed emulsions. Using this technology, Nie et al. [23] prepared monodispersed poly(tripropylene glycol diacrylate) (PTPGDA) spherical particles with surfaces armored by poly(divinylbenzene-methacrylic acid) [poly (DVB-MAA)] latex from an O/W Pickering emulsion template. TPGDA-containing oil droplets were photopolymerized after poly(DVB-MAA) latex particles migrated "inside-out" to stabilize the emulsion, which minimized fouling in the microfluidic emulsification and allowed direct visualization of droplet encapsulation.

Applications of Pickering emulsion templates in the life sciences have also emerged. For example, polymer beads bearing bacterial imprints were prepared by Ye and coworkers [24] from self-assembly of the bacteria at the O/W Pickering emulsion interface, which enabled formation of microbial recognition sites on the surface of polymer beads. This versatile bacterial recognition, based on the prepolymer and target bacteria, is promising for the construction of cell–cell communication networks, biosensors, and testing of antibiotic drugs.

In addition to solid spheres derived from simple W/O or O/W Pickering emulsions, some novel spheres with porous structures have also been prepared from more complicated Pickering double (or multiple) emulsions [25, 26]. For example, hierarchical porous polymeric microspheres (HPPMs) were fabricated from a double Pickering emulsion (W/O/W) template by Ning et al. [26]. The HPPMs were subsequently sulfonated with sulfuric acid, which endowed the pore skeleton surface with sulfonic groups, facilitating the adsorption of $[Ag(NH_3)_2]^+$. Hierarchical porous poly(styrene-$co$-DVB)–Ag nanocomposite microspheres were thus obtained by in situ reduction of $[Ag(NH_3)_2]^+$ in the presence of sodium borohydride. Such a unique porous sphere structure endowed HPPMs with outstanding performance as absorbents and catalyst scaffolds. Recently, Lei et al. [25] prepared a Pickering W/O/W high internal phase emulsion with a double emulsion morphology (HIPE-DE) and oil fraction of up to 90 vol% through a simple one-step emulsification employing freshly prepared poly[2-(diethylamino)ethyl methacrylate] (PDEA) microgel particles as the Pickering stabilizer. It was demonstrated that such W/O/W HIPE-DE could be precisely converted into either porous polystyrene (PS) spheres by loading styrene in the oil phase or into porous polyacrylamide (PAM) matrix (foam structure) by loading acrylamide in the water phase (Fig. 2).

Microspheres with very complicated morphologies have also been prepared from Pickering emulsion templates. For instance, He et al. [27] reported cage-like polymer microspheres with hollow cores and porous shells (Fig. 3). A monomer (MMA or vinyl acetate)-in-water Pickering emulsion was stabilized by surface-sulfonated PS particles, which self-assembled at the O/W droplet interface. After a certain period of time, MMA monomer in the core phase diffused into the absorbed PS particles, creating a cavity inside the core phase and swelling of the absorbed PS particles. Finally, microspheres with polymeric porous shells and hollow cores were formed by polymerizing the monomer using γ-ray irradiation.

## 2.2   Porous Capsules

Hollow capsules derived from Pickering emulsions, also known as colloidosomes, were first reported by Velev et al. [28] and named by Dinsmore et al. [6]. In contrast to hybrid latex particles, polymerization occurs only at the oil–water interface,

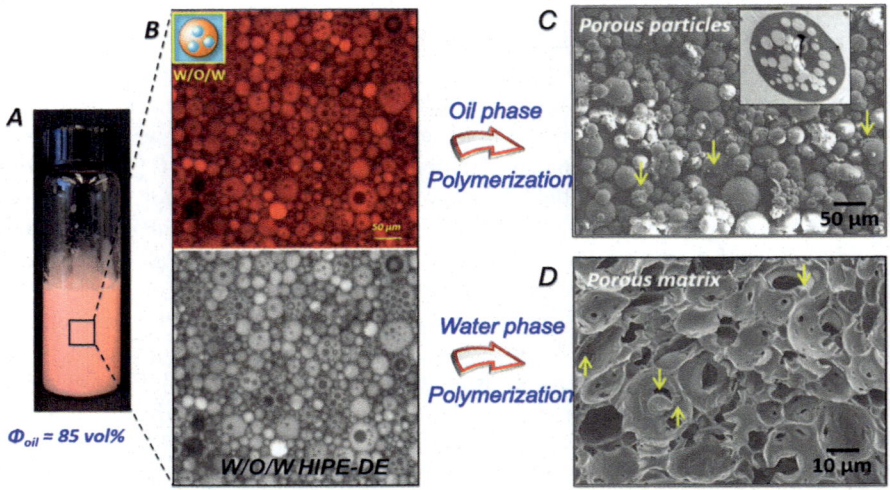

**Fig. 2** (**a**) Digital photo and (**b**) fluorescence images of Pickering HIPE-DE with 85 vol% oil phase stabilized by the readily prepared PDEA microgel (50 mg/mL). Oil phase stained with 0.01 wt% Nile red, showing *red* color in confocal fluorescence image. SEM images of (**c**) porous PS spheres prepared by polymerizing the oil-phase styrene, and (**d**) porous PAM prepared by polymerizing acrylamide dissolved in aqueous phase. Reprinted with permission from [25]

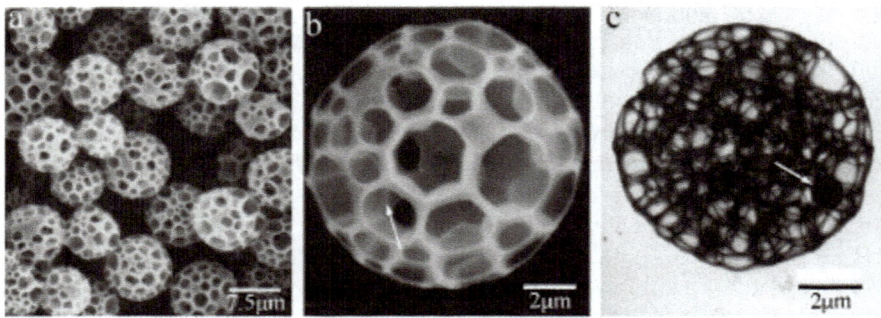

**Fig. 3** SEM (**a**, **b**) and TEM (**c**) images of PMMA cage-like microspheres with hollow core–porous shell structures. Reprinted with permission from [27]

leading to hollow spheres consisting of closely packed Pickering particles as shell [29]. Their hollow structure and semipermeability result from the interstices between particles. The functionalities of the particles promise potential applications in encapsulation [30–32], catalysis [33, 34], drug delivery [35, 36], sensors [37], and photothermal therapy [38].

Dinsmore et al. [6] described the three-stage procedure involved in fabrication of colloidosomes templated from Pickering emulsions: (1) Colloid particles self-assemble at the water–oil interface and form a Pickering emulsion.

(2) Reinforcement of the colloid shell converts the precursor Pickering emulsion droplets into robust microcapsules, which is essential for preventing disassembly of colloidosome capsules during subsequent purification and application [29]. This can be realized by binding adjacent colloid particles through thermo-annealing (sintering) [6, 39], chemical crosslinking [35, 40, 41], physical bridging by entanglement of high molecular weight polymers [42], coupling through electrostatic binding [6], and so on. (3) Purification of capsules is accomplished by removing excess colloid particles and transferring the capsules into a fresh liquid phase through centrifugation and solvent exchange. The selection of colloid particles and the binding conditions are crucial for obtaining colloidosomes with the desired properties, including size, permeability, mechanical strength, compatibility, and even functionality.

### 2.2.1 Inorganic Stabilizer

Skaff et al. first reported crosslinked capsules of quantum dots by interfacial assembly via Pickering emulsion and ligand crosslinking (Fig. 4) [43]. The quantum dots were functionalized with cyclic olefins, which were crosslinked by ring-opening metathesis polymerization after self-assembly at the oil–water interface. The carefully designed catalyst remained activity in the aqueous phase and at the interface, which was crucial for the success of polymerization.

Bon's group reported a simple and effective method for the fabrication of hollow capsules via Pickering emulsion polymerization [44]. The oil phase of crosslinkable styrene-DVB in a poor solvent underwent phase separation during polymerization. The polymer phase was driven toward the interface of colloidosomes by interfacial tensions, leading to the formation of hollow structures (Fig. 5). A silica-polymer nanocontainer with a high content of anticorrosive agent was prepared using the same procedure [45]. Bradshaw and coworkers adopted the same strategy and prepared MOF-polymer composite microcapsules [46]. They also studied the dye encapsulation and release behavior of the obtained hollow capsules and envisioned the introduction of stimuli-responsive properties for advanced applications.

Recently, a $CO_2$ Pickering emulsion template was developed by Chen and coworkers for the preparation of silica-polymer hollow capsules [47]. Silica nanoparticles were used to stabilize $CO_2$ in a water Pickering emulsion and then $CO_2$ bubbles were utilized as template for the hollow core through polymerization of melamine. The shell of the composite capsule was then transformed into a porous shell by washing, and the resulting hollow porous nitrogen-doped carbon particles showed excellent performance in lithium sulfur batteries (Fig. 6).

### 2.2.2 Organic Stabilizer

Organic particles such as latexes [28, 48, 49] and microgels [41, 50] have been widely used as building blocks in the preparation of colloidosomes. Velev et al. [28]

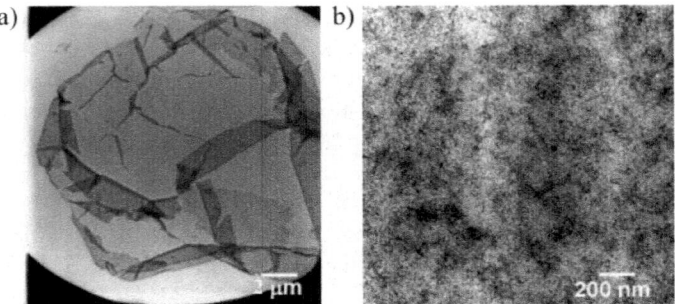

**Fig. 4** TEM images of dried quantum dot capsules at different magnifications. Reprinted with permission from [43]

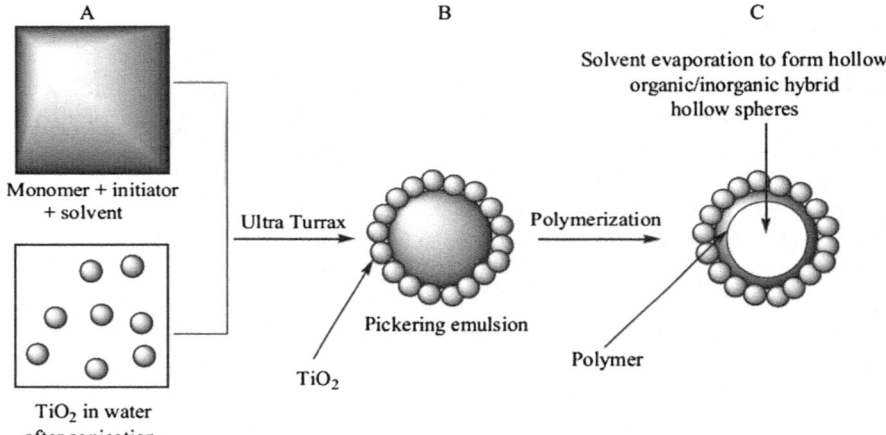

**Fig. 5** Fabrication of TiO$_2$-polymer colloidosomes: (**a**) Monomer and initiator are dissolved in *n*-hexadecane; TiO$_2$ is dispersed in water. (**b**) Formation of TiO$_2$ stabilized Pickering emulsion. (**c**) Formation of TiO$_2$-polymer colloidosomes as a result of the poor solubility of polymer in organic solvent. Reprinted with permission from [44]

reported the first example of colloidosomes prepared from surface-modified latex particles stabilized in an octanol-in-water Pickering emulsion. The size, shape, and morphology of the resulting colloidosomes could be modified by selection of appropriate substrate latexes and their compositions. This technology provides a powerful tool for developing well-defined Pickering emulsion-based microstructured and/or multicomponent materials [28, 48, 49].

Compared with rigid inorganic particles, organic particles (e.g., microgels) are richer in functionality, which provides greater opportunities for fabrication of stimuli-responsive microcapsules. The size and shell permeability of these micro-capsules can be tuned by external stimuli such as heat, pH, ionic strength, and

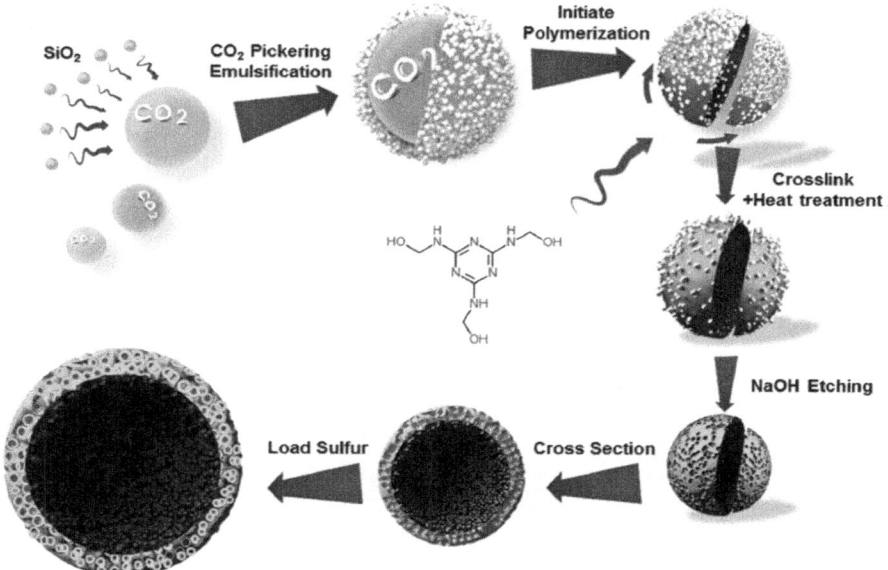

**Fig. 6** Micrometer-sized hollow porous nitrogen-doped carbon particles. Reprinted with permission from [47]

magnetic field. For example, thermoresponsive monodispersed microgel colloidosomes were prepared by Weitz and colleagues [51] based on poly(*N*-isopropylacrylamide) (PNIPAM) microgel via a microfluid Pickering emulsion. The colloidosomes exhibited reversible size variation upon environmental temperature changes, showing about 80% volume decrease when actuated. This concept is also applicable to other stimuli-responsive microcapsules by incorporating functional groups inside or on the surface of the colloid particles. Recently, Yoshida and colleagues [41] developed cell-like colloidosomes, based on thermoresponsive poly [*N*-isopropylacrylamide-*r*-(*N*-(3-aminopropyl)-methacrylamide)] [P(NIPAM-*r*-NAPMAM)] microgels, by chemically crosslinking a microgel-stabilized W/O Pickering emulsion template with P[NIPAM-*r*-(*N*-acryloxysuccinimide)] [P (NIPAM-*r*-NAS)] linear polymer dissolved in aqueous phase. The colloidosomes could undergo autonomous shape oscillations and buckling in response to changes in temperature, representing a starting point in the development of artificial cells.

Armes' group prepared various pH-responsive colloidosomes from pH-responsive latexes and microgels [52]. For example, poly(*tert*-butylaminoethyl methacrylate) latex [53] and poly(glycerol monomethacrylate)-stabilized polystyrene latex [40, 54] were used as effective O/W Pickering emulsion stabilizers for a variety of model oils. Well-defined colloidosomes were prepared via covalently crosslinking the latex particles with oil-soluble tolylene-2,4-diisocyanateterminated poly(propylene glycol) (PPG-TDI) crosslinker. Interestingly, polymer-somes self-assembled from block copolymers could also be used to stabilize O/W emulsions [55]. However, self-crosslinking of polymersomes was necessary to ensure their integrity during precursor preparation. Well-defined colloidosome microcapsules were generated by covalently crosslinking PPG-TDI. Addition of the crosslinker could raise toxicity concerns, so Wang et al. [50] used vinyl-functionalized poly(ethylacrylate-co-MMA-co-1,4-butanediol diacrylate)-glycidyl methacrylate microgels as both building block and macrocrosslinker for colloido-some preparation. Such double-crosslinked colloidosomes exhibited pH-triggered swelling in the physiological pH region, which is promising for their application in biomaterials.

Stimuli-responsive colloidosomes are promising for control-release applications. For example, Cayre et al. [35] prepared pH-switchable colloidosome microcapsules using poly(dimethylaminoethyl methacrylate) surface modified poly(methyl meth-acrylate) (PMMA) latex as building block. Both pH-triggered encapsulation and controlled release of dextran molecules were successfully demonstrated. Lei et al. (Gas-switchable microgel-colloidosome with $O_2$ and $CO_2$ tunable shell permeabil-ity for hierarchical size-selective control-release, unpublished) recently demon-strated the hierarchical loading and controlled release of cargo molecules of different molecular weights or sizes using $O_2$ and $CO_2$ dual gas-switchable poly [2-(diethylamino)ethyl methacrylate-co-2,3,4,5,6-pentafluorostyrene], P(DEA-co-FS), microgels in the preparation of microgel-colloidosomes (MGCs). The chemical compositions of the $O_2$-responsive FS and $CO_2$-responsive DEA components were crucial for the switchability of microgels, which further affected the change in shell permeability of the MGC microcapsules in response to $O_2$ and $CO_2$ treatments. The work reports the first dual gas-switchable MGCs with $O_2$- and $CO_2$-tunable shell permeability based on a size-exclusion mechanism. Such "smart" MGC systems have potential applications in the fields of medicine and health.

Besides microsphere stablizers, microrods (or microfibers) have also been used to stabilize Pickering emulsions and to develop novel colloidosomes. For example, Noble et al. [33] reported a type of "hairy" colloidosome with a shell of polymeric microrods (Fig. 7). Hot agarose aqueous solution was finely dispersed in the oil phase in the presence of rod-like particles, forming agarose gel microcapsules with gelled aqueous cores after the system was cooled. The microcapsules were purified and transferred into an aqueous phase. Compared with microsphere-based colloidosomes, these microrod-based hairy colloidosomes possess superb mecha-nical stability, which enables their use as delivery vehicles.

## 2.3 Foams

Foams, as a kind of porous matrix, can be obtained through polymerization of the continuous phase of emulsion templates, especially HIPE templates with dispersed phase ratios of 74 vol% or higher [56]. The pore size, structure, and morphology of the matrix depend on the ratio of dispersed to continuous phases, type and

**Fig. 7** Optical microscope images of "hairy" colloidosome microcapsules prepared by transferring microrod-coated agarose beads into water. Reprinted with permission from [33]

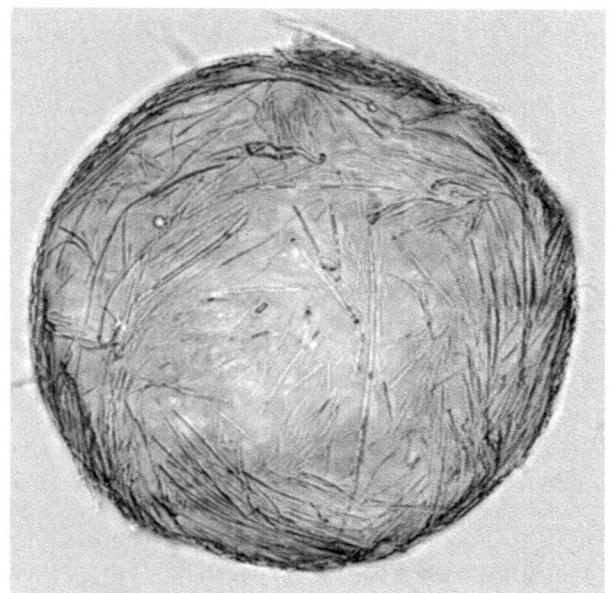

concentration of stabilizer, monomer/crosslinker composition, and so on. Based on the presence or absence of interconnecting pores (windows), foams can be classified into "open-cell" and "closed-cell" morphologies [57]. With a low dispersed phase ratio and stabilizer concentration, polymerization of the continuous phase tends to result in a closed-cell porous matrix structure after dispersed liquid is extracted. Such closed-cell structures are generally used for preparation of low density and high insulation materials. Open-cell foams can be obtained by regulating the stabilizer type, monomer content and composition, surfactant concentration, and internal phase volume fraction. However, emulsions stabilized solely by particulate stabilizers (Pickering emulsions) often form closed-cell structures after polymerization of the continuous phase, even under a high internal phase ratio [58, 59]. Poly-Pickering-HIPEs with open-cell morphology can be obtained using carefully designed systems, which are discussed in the following section.

The porous materials prepared from Pickering HIPEs often have extremely high surface areas, which facilitate applications as absorbents and catalyst carriers. Pickering stabilizers can be either organic, inorganic, or hybrid nano-/macroparticles, which can be easily surface-modified with a variety of functionalities and introduced to the cell walls of porous materials when HIPEs are polymerized. This provides opportunities for the development of advanced porous materials for use as filters, membranes, ion-exchange columns, chromatography media, chemical scavengers, and absorbents [56, 60, 63].

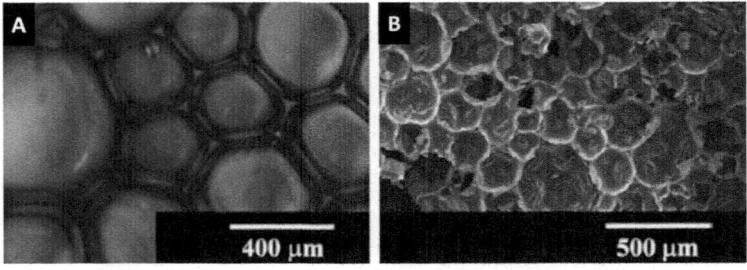

**Fig. 8** Optical microscope (**a**) and SEM (**b**) images of HIPEs with 90% internal phase volume stabilized by 4% SNPs [63]. Reprinted with permission from [63]

### 2.3.1 Inorganic Stabilizer

Early efforts were focused on the fabrication of poly(Pickering foams) through medium internal phase emulsions (MIPEs). In 2006, Bismarck and colleagues reported the use of carbon nanotube (CNT)-stabilized MIPE templates for the synthesis of porous polymer foams [61]. Although the emulsions contained only 60 vol% dispersed phase, the obtained polymer foams exhibited low densities and high degrees of pore interconnectivity. The addition of CNTs not only provided processing advantages but also enhanced the mechanical and electrical properties of the final foams. CNTs were oxidized to increase their hydrophilicity because pure CNTs could not stabilize these emulsions [62].

In 2007, Bismarck's group reported the fabrication of porous polymer foams from stable HIPE templates, solely stabilized by low concentrations of functionalized titania nanoparticles (TNPs) [56]. Commercial TNPs were modified with oleic acid to reduce their hydrophilicity in order to stabilize the O/W emulsions. HIPEs with an internal phase volume fraction of up to 0.80 were obtained. A proper balance of surface wetting characteristics of the TNPs was key in the preparation of stable HIPEs. The same group then applied this approach to silica nanoparticles (SNPs) [63]. SNPs were functionalized by oleic acid and then used as Pickering stabilizers. HIPEs with up to 92 vol% of internal phase could be stabilized by the functionalized SNPs, which were then polymerized to obtain highly porous polymer foams (Fig. 8). Magnetic macroporous polymers were also prepared using oleic acid-modified iron oxide particles stabilized Pickering HIPEs as templates [64]. The effects of dispersant addition on the morphology of poly-Pickering-HIPEs were investigated [59]. It was found that adding an oil-soluble dispersant (Hypermer 2296) changed poly-Pickering-HIPEs from closed-cell to open-cell. This was attributed to a thinning effect of the dispersant on the droplet film, making it vulnerable to breakage during polymerization or subsequent purification and drying of poly-Pickering-HIPEs.

Since then, researchers have focused on finding new methods to open the pores in poly-Pickering-HIPEs to increase their permeability. In 2014, Wang's group reported the preparation of interconnected poly-Pickering-HIPEs using modified

SNPs [65]. The pore and pore throat sizes could be controlled by the wettability and amount of particles. It was claimed that the initial location of modified silica particles significantly influenced pore throat formation. Dispersion of particles in water or oil resulted in polymer foams with open-cell or closed-cell pores, respectively. A possible reason for the formation of pore throats could be the formation of micelle-like structures of SNP aggregates in water, which were easily swollen by the monomer. Later, Wang and colleagues proposed another approach for preparation of open-cell poly-Pickering-HIPEs [66]. The films between pores were torn when sufficient volume shrinkage occurred or greater force was applied, leading to the formation of pore throats. Methyl acrylate with high volume shrinkage was used as comonomer during polymerization of styrene to open the pores and tailor the pore structure. This paper reported a facile and versatile method for preparation of interconnected poly-Pickering-HIPEs.

Besides commonly used particles such as silica and titania, other functional particles have also been reported as Pickering stabilizers for the preparation of HIPEs and their corresponding poly-Pickering-HIPEs (Fig. 9). Graphene oxide (GO) was first used by Wang and coworkers to prepare macroporous polymers via Pickering HIPE templates [67]. The effect of GO concentration on the void size of macroporous polymers was studied. The obtained polymer-GO foams were further calcinated to prepare three-dimensional macroporous chemically modified graphene. Recently, the group prepared interconnected macroporous hydrogels from GO-stabilized Pickering HIPEs [68]. By tuning the wettability and concentration of GO, an open-cell structure was obtained that showed enhanced adsorption behavior of both methylene blue and copper(II) ions. Carbonaceous microspheres were applied as Pickering stabilizers by Zhou and colleagues [69]. Closed-cell structures were obtained when these microspheres were used as the only stabilizer, whereas interconnected porous foams were prepared by introducing a small amount of a second surfactant. Poly-Pickering-HIPEs with a closed-cell structure showed high absorption capacity and absorption rate for oils, and were evaluated as absorbents for oil–water separation. Poly-Pickering-HIPEs with open-cells were highly permeable to gases as a result of their interconnected structures. Wan and coworkers used surface-modified platinum nanoparticles (PtNPs) to stabilize water-in-oil HIPEs [70]. An open-cell and elastic monolith was obtained, with PtNPs decorating the surface. The material actively catalyzed the reduction of 4-nitrophenol and exhibited good recyclability, with no reduction in catalytic activity within 20 cycles. Recently, Zhu et al. reported the use of MOF particles, a class of emerging porous crystalline materials, as stabilizers for preparation of HIPEs and the corresponding poly-Pickering-HIPEs [71]. The obtained porous materials had a closed-cell structure and were extremely light (see Fig. 9D).

### 2.3.2 Organic Stabilizer

A variety of porous foams have been prepared using Pickering emulsions stabilized by organic particles. For example, Zhang and Chen [72] prepared open porous

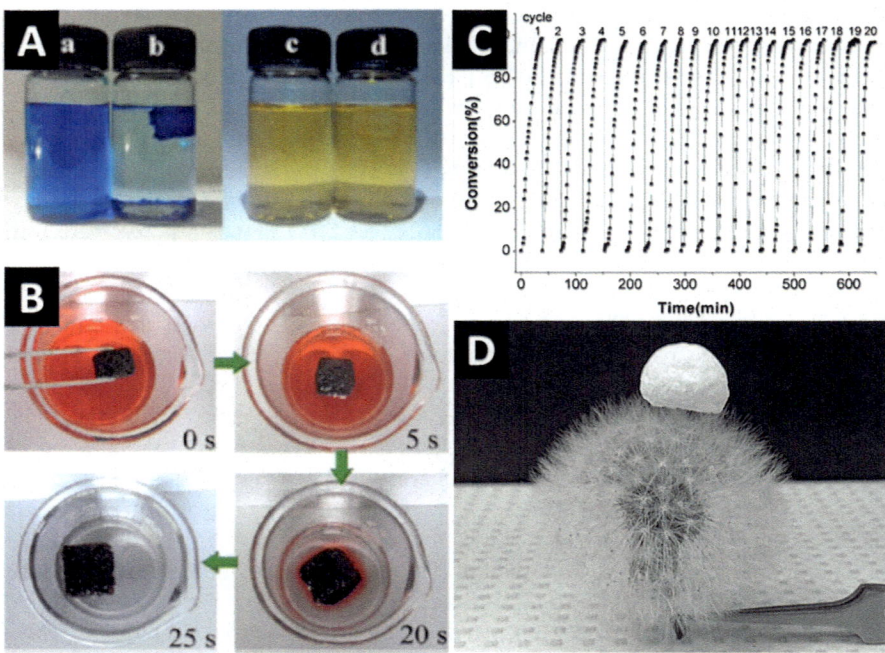

**Fig. 9** (**A**) Photographs of (**a, b**) methylene blue and (**c, d**) sunset yellow solutions (**a, c**) before and (**b, d**) after adsorption by GO–poly-Pickering-HIPEs [68]. Reprinted with permission from [68]. (**B**) Process of toluene absorption (dyed with Sudan III) from water by carbonaceous microsphere–poly-Pickering-HIPEs within 25 s [69]. Reprinted with permission from [69]. (**C**) Cyclic catalytic behavior of Pt–poly-Pickering-HIPEs for the reduction of 4-nitrophenol to 4-aminophenol [70]. Reprinted with permission from [70]. (**D**) MOF–poly-Pickering-HIPEs standing on a dandelion flower [71]. Reprinted with permission from [71]

PMMA foams based on Pickering HIPEs stabilized by P(styrene-*co*-MMA-*co*-AA) copolymer particles. A very low content of particles (1 wt%) was able to stabilize Pickering HIPEs with an internal phase ratio of up to 93.3 vol%. This work demonstrated the successful preparation of poly-HIPE foams based on intermediate hydrophilic monomers. As a byproduct of the paper industry, lignin was exploited as a particulate stabilizer for Pickering emulsions. For instance, a lignin-stabilized O/W Pickering emulsion was prepared by Wang and colleagues [58]. Interconnected macroporous foams were generated by including melamine formaldehyde prepolymer in the continuous aqueous phase and then polymerizing from a particle (lignin)–monomer (melamine formaldehyde prepolymer) coactive Pickering HIPE system.

Foam structures could also been obtained from renewable and nontoxic organic cellulose particle-stabilized Pickering emulsions. For example, Bismarck and coworkers [73] fabricated renewable macroporous cellulose nanocomposite foams from W/O HIPE stabilized by hydrophobized bacterial cellulose nanofibrils.

This work provides versatile options for the processing and application of renewable foam structures and materials. Some novel foams have been reported. For example, a flexible polymer hydrogel scaffold was prepared by Cohen and Silverstein [74]. The authors fabricated a poly(styrenesulfonate) HIPE framework filled with crosslinked poly(2-ethylhexyl acrylate) elastomer from a "one-pot" emulsion template of Pickering O/W HIPE stabilized by poly(styrenesulfonate-*co*-N,N-methylenebis(acrylamide)-*co*-styrene) hydrogel nanoparticles.

# 3 Conclusion

Pickering emulsions provide versatile templates for the fabrication of hybrid materials with different structures. Solid spheres are prepared by polymerizing the dispersed phase of an emulsion. The resulting hybrid latex particles can be used for preparation of films with novel properties. Hollow capsules or colloidosomes are usually obtained when polymerization occurs only at the oil–water interface. These capsules are promising in applications such as encapsulation, catalysis, and drug delivery. Porous foams are usually fabricated after polymerization of the continuous phase of HIPEs. These foams are promising for the development of advanced porous materials used as filters, absorbents, etc. We believe that with the development of material technology and sophisticated design of Pickering emulsion templates, these emulsions will become a powerful tool for the fabrication of advanced hybrid materials for various applications.

**Acknowledgement** The authors sincerely acknowledge the Natural Science and Engineering Research Council (NSERC) of Canada for supporting this fundamental research through the Discovery Grant program (RGPIN-2015-05841) and Canada Research Chair (950-229035) program, the Canada Foundation for Innovation (CFI) (200154) for the equipment and facilities. HZ thanks the National Natural Science Foundation of China (Grants 21420102008 and 21536011) for supporting his research.

# References

1. Ramsden W (1903) Separation of solids in the surface-layers of solutions and 'suspensions' (observations on surface-membranes, bubbles, emulsions, and mechanical coagulation). -- Preliminary account. Proc R Soc Lond 72(477–486):156–164
2. Pickering SU (1907) CXCVI.-Emulsions. J Chem Soc Trans 91(0):2001–2021
3. Binks BP (2002) Particles as surfactants—similarities and differences. Curr Opin Colloid Interface Sci 7(1–2):21–41
4. Aveyard R, Binks BP, Clint JH (2003) Emulsions stabilised solely by colloidal particles. Adv Colloid Interf Sci 100(02):503–546
5. Chevalier Y, Bolzinger MA (2013) Emulsions stabilized with solid nanoparticles: Pickering emulsions. Colloids Surf A Physicochem Eng Asp 439(2):23–34

6. Dinsmore AD, Hsu MF, Nikolaides MG, Marquez M, Bausch AR, Weitz DA (2002) Colloidosomes: selectively permeable capsules composed of colloidal particles. Science 298:1006–1009
7. Tang J, Quinlan PJ, Tam KC (2015) Stimuli-responsive Pickering emulsions: recent advances and potential applications. Soft Matter 11(18):3512–3529
8. Binks BP, Horozov TS (2006) Colloidal particles at liquid interfaces. Cambridge University Press, Cambridge
9. Li Z, Harbottle D, Pensini E, Ngai T, Richtering W, Xu Z (2015) Fundamental study of emulsions stabilized by soft and rigid particles. Langmuir 31(23):6282–6288
10. Brugger B, Vermant J, Richtering W (2010) Interfacial layers of stimuli-responsive poly-(N-isopropylacrylamide-co-methacrylicacid) (PNIPAM-co-MAA) microgels characterized by interfacial rheology and compression isotherms. Phys Chem Chem Phys 12(43):14573–14578
11. Destribats M, Lapeyre V, Wolfs M, Sellier E, Leal-Calderon F, Ravaine V, Schmitt V (2011) Soft microgels as Pickering emulsion stabilisers: role of particle deformability. Soft Matter 7(17):7689–7698
12. Ngai T, Bon S (2014) Particle-stabilized emulsions and colloids. Royal Society of Chemistry, Cambridge
13. Wu J, Ma GH (2016) Recent studies of Pickering emulsions: particles make the difference. Small 12(34):4633–4648
14. Zhang H, Su Z, Liu P, Zhang F (2007) Preparation and property of raspberry-like AS/SiO2 nanocomposite particles. J Appl Polym Sci 104(1):415–421
15. Min C, Bo Y, Zhou S, Wu L (2006) Preparation of raspberry-like PMMA/SiO2 nanocomposite particles. Front Chem Chin 1(3):340–344
16. You B, Wen N, Cao Y, Zhou S, Wu L (2009) Preparation and properties of poly[styrene-co-(butyl acrylate)-co-(acrylic acid)]/silica nanocomposite latex prepared using an acidic silica sol. Polym Int 58(5):519–529
17. Schrade A, Landfester K, Ziener U (2013) Pickering-type stabilized nanoparticles by heterophase polymerization. Chem Soc Rev 42(16):6823–6839
18. Amalvy JI, Percy MJ, Armes SP, Wiese H (2001) Synthesis and characterization of novel film-forming vinyl polymer/silica colloidal nanocomposites. Langmuir 17(16):4770–4778
19. Schmid A, Scherl P, Armes SP, Leite CAP, Galembeck F (2009) Synthesis and characterization of film-forming colloidal nanocomposite particles prepared via surfactant-free aqueous emulsion copolymerization. Macromolecules 42(11):7067–7072
20. Negrete-Herrera N, Putaux J-L, David L, Haas FD, Bourgeat-Lami E (2007) Polymer/laponite composite latexes: particle morphology, film microstructure, and properties. Macromol Rapid Commun 28(15):1567–1573
21. Wang T, Colver PJ, Bon SAF, Keddie JL (2009) Soft polymer and nano-clay supracolloidal particles in adhesives: synergistic effects on mechanical properties. Soft Matter 5(20): 3842–3849
22. González E, Bonnefond A, Barrado M, Casado Barrasa AM, Asua JM, Leiza JR (2015) Photoactive self-cleaning polymer coatings by TiO2 nanoparticle Pickering miniemulsion polymerization. Chem Eng J 281:209–217
23. Nie Z, Park JI, Li W, Bon SAF, Kumacheva E (2008) An "inside-out" microfluidic approach to monodisperse emulsions stabilized by solid particles. J Am Chem Soc 130:16508–16509
24. Shen X, Svensson Bonde J, Kamra T, Bulow L, Leo JC, Linke D, Ye L (2014) Bacterial imprinting at Pickering emulsion interfaces. Angew Chem 53(40):10687–10690
25. Lei L, Zhang Q, Shi S, Zhu S (2016) High internal phase emulsion with double emulsion morphology and their templated porous polymer systems. J Colloid Interface Sci 483:232–240
26. Ning Y, Yang Y, Wang C, Ngai T, Tong Z (2013) Hierarchical porous polymeric microspheres as efficient adsorbents and catalyst scaffolds. Chem Commun (Camb) 49(78):8761–8763
27. He X, Ge X, Liu H, Wang M, Zhang Z (2005) Synthesis of cagelike polymer microspheres with hollow core/porous shell structures by self-assembly of latex particles at the emulsion droplet interface. Chem Mater 17:5891–5892

28. Velev OD, Furusawa K, Nagayama K (1996) Assembly of latex particles by using emulsion droplets as templates. 1. Microstructured hollow spheres. Langmuir 12:2374–2384
29. Thompson KL, Williams M, Armes SP (2015) Colloidosomes: synthesis, properties and applications. J Colloid Interface Sci 447:217–228
30. Keen PH, Slater NK, Routh AF (2012) Encapsulation of lactic acid bacteria in colloidosomes. Langmuir 28(46):16007–16014
31. Keen PH, Slater NK, Routh AF (2014) Encapsulation of amylase in colloidosomes. Langmuir 30(8):1939–1948
32. Lei L, Zhang Q, Shi S, Zhu S (2016) Oxygen and carbon dioxide dual gas-switchable thermoresponsive homopolymers. ACS Macro Lett 5(7):828–832
33. Noble PF, Cayre OJ, Alargova RG, Velev OD, Paunov VN (2004) Fabrication of "hairy" colloidosomes with shells of polymeric microrods. J Am Chem Soc 126:8092–8093
34. Zhang C, Hu C, Zhao Y, Möller M, Yan K, Zhu X (2013) Encapsulation of laccase in silica colloidosomes for catalysis in organic media. Langmuir 29(49):15457–15462
35. Cayre OJ, Hitchcock J, Manga MS, Fincham S, Simoes A, Williams RA, Biggs S (2012) pH-responsive colloidosomes and their use for controlling release. Soft Matter 8(17): 4717–4724
36. Li S, Moosa BA, Croissant JG, Khashab NM (2015) Electrostatic assembly/disassembly of nanoscaled colloidosomes for light-triggered cargo release. Angew Chem 127(23):6908–6912
37. Phan-Quang GC, Lee HK, Phang IY, Ling XY (2015) Plasmonic colloidosomes as three-dimensional SERS platforms with enhanced surface area for multiphase sub-microliter toxin sensing. Angew Chem 127(33):9827–9831
38. Liu D, Zhou F, Li C, Zhang T, Zhang H, Cai W, Li Y (2015) Black gold: plasmonic colloidosomes with broadband absorption self-assembled from monodispersed gold nanospheres by using a reverse emulsion system. Angew Chem Int Ed 54(33):9596–9600
39. Cayre OJ, Noble PF, Paunov VN (2004) Fabrication of novel colloidosome microcapsules with gelled aqueous cores. J Mater Chem 14(22):3351–3355
40. Thompson KL, Armes SP, Howse JR, Ebbens S, Ahmad I, Zaidi JH, York DW, Burdis JA (2010) Covalently cross-linked colloidosomes. Macromolecules 43(24):10466–10474
41. Tamate R, Ueki T, Yoshida R (2016) Evolved colloidosomes undergoing cell-like autonomous shape oscillations with buckling. Angew Chem 55(17):5179–5183
42. Berger S, Zhang H, Pich A (2009) Microgel-based stimuli-responsive capsules. Adv Funct Mater 19(4):554–559
43. Skaff H, Lin Y, Tangirala R, Breitenkamp K, Böker A, Russell TP, Emrick T (2005) Crosslinked capsules of quantum dots by interfacial assembly and ligand crosslinking. Adv Mater 17(17):2082–2086
44. Chen T, Colver PJ, Bon SAF (2007) Organic–inorganic hybrid hollow spheres prepared from TiO2-stabilized Pickering emulsion polymerization. Adv Mater 19(17):2286–2289
45. Haase MF, Grigoriev DO, Möhwald H, Shchukin DG (2012) Development of nanoparticle stabilized polymer nanocontainers with high content of the encapsulated active agent and their application in water-borne anticorrosive coatings. Adv Mater 24(18):2429–2435
46. Huo J, Marcello M, Garai A, Bradshaw D (2013) MOF-polymer composite microcapsules derived from Pickering emulsions. Adv Mater 25(19):2717–2722
47. Li M, Zhang Y, Wang X, Ahn W, Jiang G, Feng K, Lui G, Chen Z (2016) Gas Pickering emulsion templated hollow carbon for high rate performance lithium sulfur batteries. Adv Funct Mater 26:8408–8417
48. Velev OD, Nagayama K (1997) Assembly of latex particles by using emulsion droplets. 3. Reverse (water in oil) system. Langmuir 13:1856–1859
49. Velev OD, Furusawa K, Nagayama K (1996) Assembly of latex particles by using emulsion droplets as templates. 2. Ball-like and composite aggregates. Langmuir 12:2385–2391
50. Wang W, Milani AH, Carney L, Yan J, Cui Z, Thaiboonrod S, Saunders BR (2015) Doubly crosslinked microgel-colloidosomes: a versatile method for pH-responsive capsule assembly using microgels as macro-crosslinkers. Chem Commun 51(18):3854–3857

51. Shah RK, Kim JW, Weitz DA (2010) Monodisperse stimuli-responsive colloidosomes by self-assembly of microgels in droplets. Langmuir 26(3):1561–1565
52. Yuan Q, Cayre OJ, Fujii S, Armes SP, Williams RA, Biggs S (2010) Responsive core-shell latex particles as colloidosome microcapsule membranes. Langmuir 26(23):18408–18414
53. Morse AJ, Madsen J, Growney DJ, Armes SP, Mills P, Swart R (2014) Microgel colloidosomes based on pH-responsive poly(tert-butylaminoethyl methacrylate) latexes. Langmuir 30(42):12509–12519
54. Walsh A, Thompson KL, Armes SP, York DW (2010) Polyamine-functional sterically stabilized latexes for covalently cross-linkable colloidosomes. Langmuir 26(23):18039–18048
55. Thompson KL, Chambon P, Verber R, Armes SP (2012) Can polymersomes form colloidosomes? J Am Chem Soc 134(30):12450–12453
56. Menner A, Ikem V, Salgueiro M, Shaffer MSP, Bismarck A (2007) High internal phase emulsion templates solely stabilised by functionalised titania nanoparticles. Chem Commun 43(41): 4274–4276
57. Zhang T, Xu Z, Guo Q (2016) Closed-cell and open-cell porous polymers from ionomer-stabilized high internal phase emulsions. Polym Chem 7(48):7469–7476
58. Yang Y, Wei Z, Wang C, Tong Z (2013) Lignin-based Pickering HIPEs for macroporous foams and their enhanced adsorption of copper(II) ions. Chem Commun 49(64):7144–7146
59. Ikem VO, Menner A, Horozov TS, Bismarck A (2010) Highly permeable macroporous polymers synthesized from Pickering medium and high internal phase emulsion templates. Adv Mater 22(32):3588–3592
60. Barbetta A, Cameron NR (2004) Morphology and surface area of emulsion-derived (PolyHIPE) solid foams prepared with oil-phase soluble porogenic solvents: three-component surfactant system. Macromolecules 37(9):3188–3201
61. Menner A, Verdejo R, Shaffer M, Bismarck A (2007) Particle-stabilized surfactant-free medium internal phase emulsions as templates for porous nanocomposite materials: poly-Pickering-foams. Langmuir 23(5):2398–2403
62. Hermant MC, Verhulst M, Kyrylyuk AV, Klumperman B, Koning CE (2009) The incorporation of single-walled carbon nanotubes into polymerized high internal phase emulsions to create conductive foams with a low percolation threshold. Compos Sci Technol 69(5):656–662
63. Ikem VO, Menner A, Bismarck A (2008) High internal phase emulsions stabilized solely by functionalized silica particles. Angew Chem Int Ed 47(43):8277–8279
64. Vílchez A, Rodríguezabreu C, Esquena J, Menner A, Bismarck A (2011) Macroporous polymers obtained in highly concentrated emulsions stabilized solely with magnetic nanoparticles. Langmuir 27(27):13342–13352
65. Zheng X, Zhang Y, Wang H, Du Q (2014) Interconnected macroporous polymers synthesized from silica particle stabilized high internal phase emulsions. Macromolecules 47(19): 6847–6855
66. Xu H, Zheng X, Huang Y, Wang H, Du Q (2016) Interconnected porous polymers with tunable pore throat size prepared via Pickering high internal phase emulsions. Langmuir 32(1):38–45
67. Zheng Z, Zheng X, Wang H, Du Q (2013) Macroporous graphene oxide-polymer composite prepared through Pickering high internal phase emulsions. ACS Appl Mater Interfaces 5(16): 7974
68. Yi W, Wu H, Wang H, Du Q (2016) Interconnectivity of macroporous hydrogels prepared via graphene oxide-stabilized Pickering high internal phase emulsions. Langmuir 32(4):982–990
69. Yu S, Tan H, Wang J, Liu X, Zhou K (2015) High porosity supermacroporous polystyrene materials with excellent oil–water separation and gas permeability properties. ACS Appl Mater Interfaces 7(12):6745–6753
70. Liu H, Wan D, Du J, Jin M (2015) Dendritic amphiphile mediated one-pot preparation of 3D Pt nanoparticles-decorated PolyHIPE as a durable and well-recyclable catalyst. ACS Appl Mater Interfaces 7(37):20885–20892

71. Zhu H, Zhang Q, Zhu S (2016) Assembly of a metal–organic framework into 3 D hierarchical porous monoliths using a Pickering high internal phase emulsion template. Chem Eur J 22(26): 8751–8755

72. Zhang S, Chen J (2009) PMMA based foams made via surfactant-free high internal phase emulsion templates. Chem Commun (16):2217–2219

73. Blaker JJ, Lee K-Y, Li X, Menner A, Bismarck A (2009) Renewable nanocomposite polymer foams synthesized from Pickering emulsion templates. Green Chem 11(9):1321

74. Cohen N, Silverstein MS (2012) One-pot emulsion-templated synthesis of an elastomer-filled hydrogel framework. Macromolecules 45(3):1612–1621

Adv Polym Sci (2018) 280: 121–194
DOI: 10.1007/12_2017_16
© Springer International Publishing AG 2017
Published online: 12 August 2017

# Modeling of Suspension Vinyl Chloride Polymerization: From Kinetics to Particle Size Distribution and PVC Grain Morphology

**Costas Kiparissides**

**Abstract** A comprehensive multiscale, multiphase modeling approach is developed to describe the dynamic evolution of polymerization rate, average molecular weight, and morphological properties of poly(vinyl chloride) (PVC) produced in batch suspension polymerization reactors. Dynamic evolution of the molecular (molecular weight distribution, long chain branching, short chain branching, terminal double bonds) and morphological (particle size distribution, grain porosity) properties of PVC can be calculated from the numerical solution of the proposed integrated model. In particular, polymer molecular properties are determined by employing a detailed kinetic mechanism that describes the free-radical polymerization of vinyl chloride monomer in both monomer- and polymer-rich phases. The initial monomer droplet size distribution and final polymer particle size distribution depend on the type and concentration of the surface-active agents, the quality of agitation (reactor geometry, impeller type, power input, etc.) and the physical properties (density, viscosity, interfacial tension, etc.) of the continuous and dispersed phases. A dynamic discretized particle population balance equation (PBE) is numerically solved to calculate the dynamic evolution of the particle size distribution of the produced PVC in a batch suspension reactor. Furthermore, the primary particle size distribution inside the polymerizing monomer droplets, which affects the porosity of the final PVC grains, is determined from the solution of a PBE governing the nucleation, growth, and aggregation of primary particles inside the polymerizing monomer droplets. Theoretical model predictions are compared

C. Kiparissides (✉)
Department of Chemical Engineering, Aristotle University of Thessaloniki, Thessaloniki, Greece

Chemical Process & Energy Resources Institute, CERTH, PO Box 60361, 57001 Thessaloniki, Greece
e-mail: costas.kiparissides@cperi.certh.gr

successfully with a comprehensive series of experimental data on polymerization
kinetics, particle size distribution, and PVC grain morphology.

**Keywords** Grain morphology • Molecular weight distribution • Multiphase
modeling • Multiscale • Particle size distribution • Primary particle size
distribution • Suspension PVC process

## Contents

## Abbreviations

| | |
|---|---|
| DH | Degree of hydrolysis |
| DSD | Droplet size distribution |
| HCl | Hydrogen chloride |
| HPMC | Hydroxypropyl methylcellulose |
| LCB | Long chain branch |
| LP40 | Lauroyl peroxide |
| LUP610 | 3-Hydroxy-1,1-dimethylbutyl peroxyneodecanoate |
| PBE | Population balance equation |

| PDEH  | Di (2-ethylhexyl) peroxydicarbonate          |
|-------|----------------------------------------------|
| PPSD  | Primary particle size distribution           |
| PSD   | Particle size distribution                   |
| PVA   | Poly(vinyl alcohol)                          |
| PVC   | Poly(vinyl chloride)                         |
| SCB   | Short chain branch                           |
| SEM   | Scanning electron microscope                 |
| TDB   | Terminal double bond                         |
| TNCLD | Total number chain length distribution       |
| VCM   | Vinyl chloride monomer                       |

# 1  Introduction

Poly(vinyl chloride) (PVC) production is, by volume, the second largest in the world for thermoplastics [1]. Global consumption of PVC in 2013 was estimated to be approximately 39 million tons. The global demand for PVC is expected to increase by about 3.2% per year until 2021. The sustainable expansion of the PVC industry is a result of the high versatility of PVC as a plastic raw material, together with its low price. A review of the qualitative and quantitative aspects of PVC polymerization can be found in the literature [1–8].

Four polymerization processes (i.e., suspension, bulk, emulsion, and solution) are commercially employed for PVC manufacture. Approximately 80% of the total PVC production is obtained by the suspension polymerization process. According to this process, droplets of liquid vinyl chloride monomer (VCM), containing oil-soluble initiator(s), are dispersed in the continuous aqueous phase by a combination of stirring and the use of suspending agents (stabilizers). The reaction takes place inside the monomer droplets. The polymerization is carried out in large batch reactors (e.g., 150 m$^3$; see Fig. 1). The reactor content is heated to the required temperature, at which the initiator(s) start(s) decomposing to produce primary free radicals. The polymerization reaction is strongly exothermic (i.e., 100 kJ mol$^{-1}$). Thus, efficient removal of reaction heat is crucial for the operation of large-scale reactors [9]. Polymerization heat is transferred from the monomer droplets to the aqueous phase and then to the reactor wall, which is cooled by chilled water flowing through the jacket of the reactor. In large-scale reactors, overhead condensers remove part of the reaction heat by monomer evaporation and subsequent condensation. When all of the free liquid monomer has been used up, the pressure in the reactor starts to fall as a result of monomer mass transfer from the vapor phase to the polymer phase due to sub-saturation conditions. In industrial PVC production, the reaction is usually stopped when a certain pressure drop has been recorded. Because the polymer is effectively insoluble in its own monomer, once the polymer chains are first generated, they precipitate immediately to form a separate phase in the polymerizing droplets. Thus, from a kinetic point of view, the polymerization of VCM is considered to take place in three stages [10].

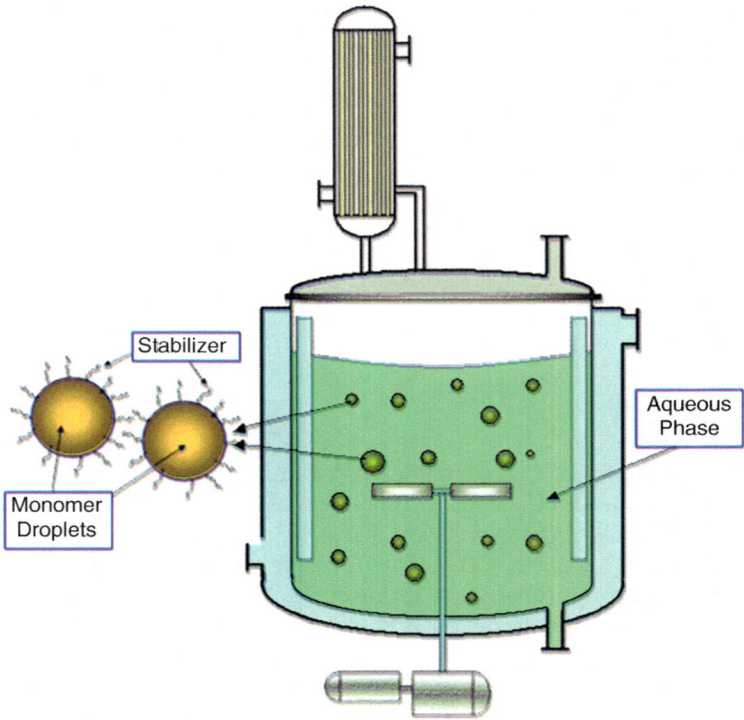

**Fig. 1** Suspension polymerization process

During the first stage, primary radicals formed by thermal fragmentation of the initiator molecules rapidly react with monomer to produce the first polymer chains. During this early polymerization period, the polymer concentration is below its solubility limit in the monomer (i.e., VCM conversion is less than 0.1%) and, therefore, the polymerization occurs in a single homogeneous phase.

The second polymerization stage extends from the time of appearance of a separate polymer phase, in addition to the monomer phase, up to a fractional monomer conversion, $X_f$, at which the separate monomer phase disappears. During this stage, the reaction mixture consists of four phases, namely, the monomer-rich and polymer-rich phases inside the polymerizing monomer droplets, the continuous aqueous phase, and the gas phase. The reaction takes place in the monomer and polymer phases at different rates and is accompanied by transfer of monomer from the monomer phase to the polymer phase so that the latter remains saturated with monomer. Disappearance of the monomer phase is associated with a characteristic drop in reactor pressure.

In the third stage, at higher monomer conversions ($X_f < X < 1.0$), the polymerization proceeds exclusively in the polymer-rich phase, which is swollen with residual monomer. Thus, the monomer mass fraction in the polymer phase continuously decreases as the total monomer conversion approaches its final limiting value.

In the last 45 years, several kinetic models have been developed to describe the multiphase suspension polymerization of VCM [7, 9–24]. However, due to the complexity of physical and chemical phenomena taking place in the reactor (reactor kinetics, phase equilibrium, particle size distribution, etc.), there are only a limited number of papers dealing with the development of comprehensive, quantitative models describing the various phenomena occurring at different modeling scales. Kiparissides et al. [23–25] have described the dynamic behavior of suspension VCM polymerization at the laboratory, pilot, and industrial scales, taking into account the complex physical and chemical phenomena occurring during VCM suspension polymerization. In most kinetic studies, a single initiator and/or mixtures of monofunctional initiators have been considered. However, there is a growing interest in the use of multifunctional initiators [24]. Multifunctional initiators usually contain two labile groups having different thermal decomposition characteristics. These initiators follow a completely different decomposition mechanism from the well-known mechanism of monofunctional initiators. As a result, substantially higher polymerization rates can be achieved, with concomitant reduction in batch time.

In the present review, a comprehensive multiscale, multiphase mathematical model is developed to describe the dynamic evolution of polymerization rate, number-average molecular weight ($M_n$), weight-average molecular weight ($M_w$), and morphological properties (i.e., particle size distribution and grain porosity) of PVC produced in batch suspension polymerization reactors. In particular, a kinetic model is developed based on a detailed mechanism for free-radical polymerization of VCM in the presence of a mixture of monofunctional initiators. Accordingly, general population balance equations (PBEs) are derived to describe the dynamic evolution of "live" and "dead" polymer chains in the monomer-rich and polymer-rich phases. The method of moments is invoked to recast the infinite set of dynamic polymer chain conservation equations into a lower system of ordinary differential equations for the leading moments of live and dead number chain length distributions. These are then used to calculate the dynamic evolution of average molecular properties ($M_n$ and $M_w$, long chain branching, short chain branching, etc.) in a suspension PVC reactor. Thermodynamic equilibrium equations are derived for calculation of monomer distribution in the different phases and prediction of time variation in reactor pressure [23, 24]. A monomer mass balance equation is also derived, taking into account monomer partitioning in the four phases, to calculate the time evolution of monomer conversion and polymerization rate during the whole course of polymerization. Gel-, glass-, and cage-effect phenomena regarding the termination and propagation rate coefficients, as well as the time-varying initiator efficiency, are accounted for using the fundamental model of Xie et al. [7]. Finally, dynamic energy balance equations for the reaction mixture, coolant in the jacket, and operation of the overhead condenser are derived to calculate the reaction temperature, inlet and outlet temperatures, flow rates of coolants into the reactor jacket and condenser, and the temperature of the reactor metal wall.

Subsequently, a generalized population balance model is developed to describe the dynamic evolution of monomer and particle size distributions (PSD) [25]. The model takes into account dynamic evolution of the physical and transport properties of the continuous and dispersed phases, in terms of monomer conversion, type and concentration of suspending agents, and the turbulent intensity characteristics of the flow field.

Finally, a population balance model is developed to describe the dynamic evolution of the primary particle size distribution (PPSD) inside the polymerizing monomer droplets [26]. The PPSD is controlled by many process variables, including polymerization temperature, ionic strength of the medium, and the type and concentration of secondary stabilizer. Its dynamic evolution is determined by the solution of a PBE governing the nucleation, growth, and aggregation of primary particles. The porosity of PVC grains is a consequence of the heterophase nature of the bulk/suspension polymerization of VCM caused by the multistage agglomeration of primary particles formed directly during polymerization. Porosity is one of the most fundamental quantities and is directly related to the internal structure of suspension PVC resins. It represents the volume fraction of PVC grains corresponding to the interior void of particle pores. In the present review, a simple porosity model is derived in terms of the critical monomer conversion ($X_c$) that describes the evolution of grain porosity with respect to monomer conversion.

## 2   Calculation of Polymerization Rate and Molecular Properties

A comprehensive kinetic mechanism is proposed for the free-radical polymerization of vinyl chloride. Based on the postulated kinetic mechanism, general rate functions for the production of live and dead polymer chains in the two reaction phases (monomer-rich and polymer-rich) are derived. The method of moments is employed to follow molecular weight developments in the two-phase heterogeneous polymerization system. Moreover, three additional dynamic balances are derived to account for the structural characteristics (number of short and long chain branches and number of double bonds) of the polymer. To account for diffusion-controlled reactions (i.e., gel-, glass-, and cage-effects) in VCM heterogeneous polymerization, a comprehensive model based on the free-volume theory is employed [7]. The concentrations of VCM in the four phases (monomer-rich, polymer-rich, aqueous, and gas) are calculated by assuming that the four phases are in thermodynamic equilibrium [23, 24]. Most of the kinetic and thermodynamic parameters in the model were taken from the work of Kiparissides et al. [23, 24] or/and estimated by fitting model predictions to a comprehensive set of experimental data on VCM conversion, polymerization rate, reactor pressure, and average molecular properties of PVC.

## 2.1    Kinetic Mechanism of VCM Free-Radical Polymerization

In general, the free-radical polymerization of vinyl monomers includes chain initiation, propagation, chain transfer to monomer, and bimolecular termination reactions. However, there is strong evidence that, in the free-radical polymerization of VCM, some reactions (e.g., chain transfer to monomer, formation of short and long chain branches) involve very complex kinetic mechanisms. The presence of chloromethyl and ethyl short chain branches in PVC corroborates the idea that propagation reactions involve several types of radicals [27–31]. Figures 2, 3, and 4, respectively, show in detail the mechanisms leading to the formation of chloromethyl and ethyl branches and terminal double bonds (TDBs), long chain branches (LCBs) and internal double bonds, and short chain branches (SCBs) via a backbiting reaction mechanism.

The work of Starnes et al. [27, 28] describes how different types of SCBs can be produced via backbiting reactions. The 1,3-diethyl branch structure is always produced at low concentrations and can therefore be ignored. Furthermore, the 1,6-shift backbiting reaction is ordinarily slower than the analogous 1,5-shift backbiting reaction. Thus, a single backbiting reaction can be employed to account for the formation of dichlorobutyl branches.

To calculate the polymerization rate and the main molecular features of the PVC chains (e.g., $M_n$ and $M_w$, SCBs, LCBs, amount of double bonds) [7, 23, 24, 32, 33], we can use the following kinetic mechanism that describes the free-radical polymerization of VCM initiated by a mixture of monofunctional initiators.

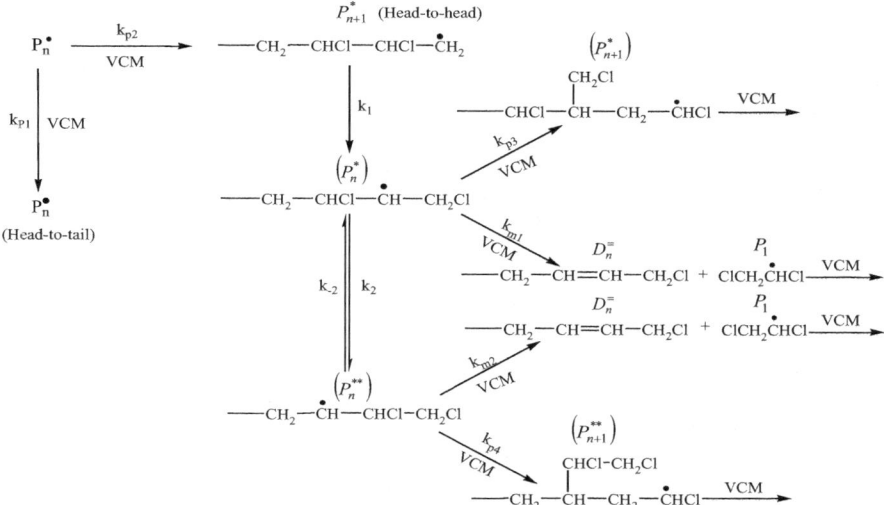

**Fig. 2** Formation of chloromethyl and ethyl branches and terminal double bonds [27]

P'm+1
CH2—CHCl
—CHCl-CH—CHCl—
VCM
Pn
Pm
kh
Dm
kb''
P'm
kp5
VCM
Dn + —CHCl-CH-CHCl—
km3
VCM
Dm
P1
—CH=CHCHCl— + ClCH2CHCl
VCM

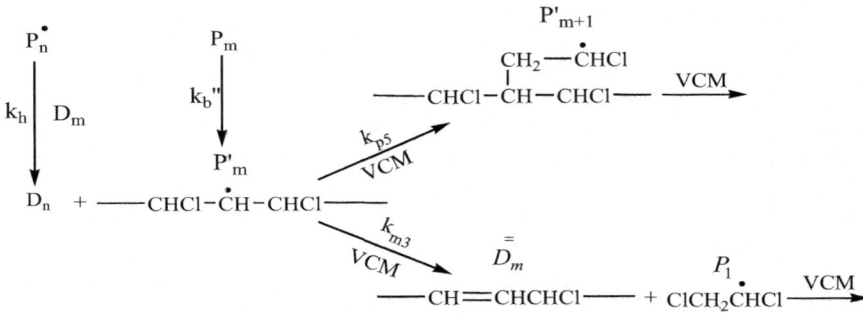

**Fig. 3** Formation of long chain branches and internal double bonds [27]

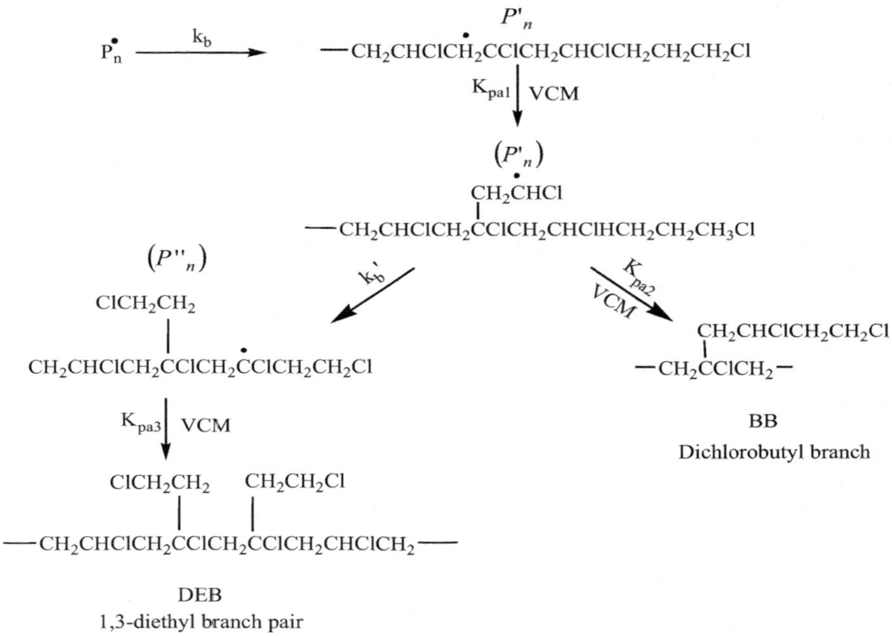

P'n
Pn
kb
—CH2CHClCH2CClCH2CHClCH2CH2CH2Cl
Kpa1
VCM
(P'n)
CH2CHCl
—CH2CHClCH2CClCH2CHClHCH2CH2CH3Cl
(P''n)
kb'
Kpa2
VCM
ClCH2CH2
CH2CHClCH2CClCH2CClCH2CH2Cl
CH2CHClCH2CH2Cl
—CH2CClCH2—
BB
Dichlorobutyl branch
Kpa3
VCM
ClCH2CH2
CH2CH2Cl
—CH2CHClCH2CClCH2CClCH2CHClCH2—
DEB
1,3-diethyl branch pair

**Fig. 4** Formation of short chain branches [27]

## Initiation:

$$I_{i,j} \overset{k_{d_{i,j}}}{\rightarrow} 2I^{\bullet}_{i,j}; \qquad i = 1, 2, \ldots, N_d \tag{1}$$

$$I^{\bullet}_{i,j} + M_j \overset{k_{I_{i,j}}}{\rightarrow} R^{\bullet}_{1,j} \tag{2}$$

**Propagation:**

$$R_{x,j}^{\bullet} + M_j \xrightarrow{k_{p_j}} R_{x+1,j}^{\bullet} \tag{3}$$

**Chain transfer to monomer:**

$$R_{x,j}^{\bullet} + M_j \xrightarrow{k_{fm_j}} P_x + R_{1,j}^{\bullet} \tag{4}$$

**Chain transfer to polymer:**

$$R_{x,2}^{\bullet} + P_y \xrightarrow{k_{fp_2}} P_x + R_{y,2}^{\bullet} \tag{5}$$

**Intramolecular transfer (backbiting):**

$$R_{x,j}^{\bullet} \xrightarrow{k_{b_j}} R_{x,j}^{\bullet} \tag{6}$$

**Termination by disproportionation:**

$$R_{x,j}^{\bullet} + R_{y,j}^{\bullet} \xrightarrow{k_{td_j}} P_x + P_y \tag{7}$$

**Termination by combination:**

$$R_{x,j}^{\bullet} + R_{y,j}^{\bullet} \xrightarrow{k_{tc_j}} P_{x+y} \tag{8}$$

**Inhibition:**

$$R_{x,j}^{\bullet} + Z_j \xrightarrow{k_{Z_j}} P_x + Z_j^{\bullet} \tag{9}$$

$$Z_j^{\bullet} + Z_j^{\bullet} \xrightarrow{k_{Z_{t_j}}} Z_j + \text{Inactive Products} \tag{10}$$

In the above kinetic scheme, the symbols $I$, $Z$, and $M$ denote the initiator, inhibitor, and monomer molecules, respectively. Radicals formed via the thermal initiator decomposition and the inhibition reaction, are denoted by the symbols $I^{\bullet}$, and $Z^{\bullet}$, respectively. $N_d$ is the number of initiators used in the polymerization. The symbols $R_x^{\bullet}$ and $P_x$ identify the respective live and dead polymer chains, containing $x$ monomer units. All other symbols are defined in the list at the end of the chapter.

The main raw material used in the production of PVC is VCM. Controlling the purity of VCM is essential in controlling the parameters of the polymerization reaction and the properties of PVC resins, such as the molecular weight, porosity, particle size, and thermal stability. VCM contaminants, including aliphatic and aromatic hydrocarbons, organochlorines, alcohols, and phenols, have been identified in the various stages of VCM production, storage, and recovery [34]. Contaminants can increase the reaction time or/and lower the VCM conversion as a result of radical inhibition. Moreover, they can react as co-monomers or chain transfer agents and, thus, their concentrations should be closely controlled.

Based on the general kinetic scheme considered in this study, Eqs. (1)–(10), one can identify three additional structural characteristics of the polymer chains related to the number of LCBs, SCBs, and TDBs. It should be noted that all the above elementary reactions, except chain transfer to polymer, can take place either in the monomer phase ($j = 1$) or/and in the polymer phase ($j = 2$).

## 2.2 Moment Rate Functions

To simplify derivation of the dynamic molar balance equations describing the conservation of the various live and dead polymer chains, the following assumptions are made:

1. Polymerization of VCM in the water and vapor phases is negligible
2. Polymerization of VCM proceeds in one phase (monomer-rich) when conversion is less than 0.1%, in two phases (monomer-rich and polymer-rich phases) in the conversion range $0.1 < X < X_f$, and in one phase (polymer-rich) at higher conversions $X > X_f$
3. No transfer of radicals between the two phases occurs
4. It is assumed that the partitioning of monomer, initiator(s), and inhibitor in the monomer-rich and the polymer-rich phases is at equilibrium at all times
5. All the kinetic rate coefficients are independent of the polymer chain length
6. The quasi-steady-state approximation is only applied to the initiator primary radicals

Accordingly, based on the postulated kinetic mechanism and assumptions, the following general rate functions for the net production of live and dead polymer chains can be derived:

$$r_{R^\bullet_{x,j}} = \left( \sum_{i=1}^{N_d} 2f_{i,j} k_{d_{i,j}} [I]_{i,j} + k_{fm_j} [M]_j \sum_{y=1}^{\infty} \left[ R^\bullet_{y,j} \right] \right) \varphi_j \delta(x-1)$$
$$+ k_{p_j} [M]_j \left[ R^\bullet_{x-1,j} \right] \varphi_j [1 - \delta(x-1)]$$
$$- \left\{ \left( k_{p_j} + k_{fm_j} \right) [M]_j + k_{Z_j} [Z]_j + (k_{tc_j} + k_{td_j}) \sum_{y=1}^{\infty} \left[ R^\bullet_{y,j} \right] \right\} \left[ R^\bullet_{x,j} \right] \varphi_j \tag{11}$$
$$+ \left( k_{fp_2} x [P_x] \sum_{y=1}^{\infty} \left[ R^\bullet_{y,j} \right] - k_{fp_2} \left[ R^\bullet_{x,j} \right] \sum_{y=1}^{\infty} y [P_y] \right) \varphi_2 (j-1)$$

$$r_{P_x} = \sum_{j=1}^{2} \left( k_{fm_j} [M]_j + k_{Z_j} [Z]_j + k_{td_j} \sum_{y=1}^{\infty} \left[ R^\bullet_{y,j} \right] \right) \varphi_j \left[ R^\bullet_{x,j} \right]$$
$$+ \frac{1}{2} \sum_{j=1}^{2} k_{tc_j} \sum_{y=1}^{x-1} \left[ R^\bullet_{y,j} \right] \left[ R^\bullet_{x-y,j} \right] \varphi_j - \left( k_{fp_2} x [P_x] \sum_{y=1}^{\infty} \left[ R^\bullet_{y,j} \right] - k_{fp_2} \left[ R^\bullet_{x,j} \right] \sum_{y=1}^{\infty} y [P_y] \right) \varphi_2 \tag{12}$$

$\delta(x)$ is Kronecker's delta and is given by:

$$\delta(x) = \begin{cases} 1 & \text{if } x = 0 \\ 0 & \text{if } x \neq 0 \end{cases} \tag{13}$$

Note that $\varphi_1$ and $\varphi_2$ refer to the volume fractions of the monomer-rich phase ($V_1/V$ for $j = 1$) and polymer-rich phase ($V_2/V$ for $j = 2$), respectively. $V$ is the total volume of the polymer particles. To reduce the infinite system of molar balance equations required to describe the time evolution of the molecular weight distribution, the method of moments is invoked. Accordingly, the average molecular properties of the polymer (e.g., $M_n$, $M_w$) are expressed in terms of the leading moments of the total number chain length (TNCL) distribution of the dead chains. Note that the respective moments of live and dead TNCL distributions are defined as:

$$\lambda_{i,j} = \sum_{x=1}^{\infty} x^i R^\bullet_{x,j}; \quad \mu_i = \sum_{x=1}^{\infty} x^i P_x; \quad i = 0, 1, 2, \ldots \tag{14}$$

The corresponding moment rate functions can be obtained by multiplying each term in Eqs. (11) and (12) by $x^i$ and summing the resulting expressions over the total range of variation of $x$ [35–37].

## Moment equations for live polymer chains:

$$
\begin{aligned}
r_{\lambda_{i,j}} &= \sum_{k=1}^{N_d} 2f_{k,j} k_{d_{k,j}} [I]_{k,j} \varphi_j + k_{\mathrm{fm}_j} [M]_j [\lambda_{0,j}] \varphi_j + k_{p_j} [M]_j \left\{ \sum_{k=0}^{i} \binom{i}{k} [\lambda_{k,j}] \right\} \varphi_j \\
&\quad - \left\{ \left( k_{p_j} + k_{\mathrm{fm}_j} \right) [M]_j + k_{Z_j} [Z]_j + \left( k_{\mathrm{tc}_j} + k_{\mathrm{td}_j} \right) [\lambda_{0,j}] \right\} [\lambda_{i,j}] \varphi_j \\
&\quad + \left( k_{\mathrm{fp}_2} [\lambda_{0,j}] [\mu_{i+1}] - k_{\mathrm{fp}_2} [\lambda_{i,j}] [\mu_1] \right) \varphi_2 (j-1)
\end{aligned}
\tag{15}
$$

## Moment equations for dead polymer chains:

$$
\begin{aligned}
r_{\mu_i} &= \sum_{j=1}^{2} \left( k_{\mathrm{fm}_j} [M]_j + k_{Z_j} [Z]_j + k_{\mathrm{td}_j} [\lambda_{0,j}] \right) [\lambda_{i,j}] \varphi_j + \frac{1}{2} \sum_{j=1}^{2} k_{\mathrm{tc}_j} \sum_{k=0}^{i} \binom{i}{k} [\lambda_{k,j}] [\lambda_{i-k,j}] \varphi_j \\
&\quad - \left( k_{\mathrm{fp}_2} [\lambda_{0,j}] [\mu_{i+1}] - k_{\mathrm{fp}_2} [\lambda_{i,j}] [\mu_1] \right) \varphi_2
\end{aligned}
\tag{16}
$$

It should be pointed out that when transfer to polymer reactions are included in the kinetic mechanism, the $i$-order polymer moment equation depends on the $(i+1)$-order moment. This is because the rate function for the transfer to polymer reaction depends on the total number of monomer units in the polymer chains. Several closure methods have been proposed for breaking down the dependence of moment equations on higher order moments. The present investigation employs the method of the so-called bulk moments. According to this closure method [38], a "bulk moment" ($\mu_i^b = \mu_i + \lambda_{i,1} + \lambda_{i,2}$) can be defined that includes the contributions of both dead and live polymer chains. Notice that the bulk moment ($\mu_i^b$) can be approximated by $\mu_i$ because of the relatively small contributions of $\lambda_{i,1}$ and $\lambda_{i,2}$. Thus, by adding the second-order live radical moment equations to the second-order dead polymer moment equation, one can obtain the following expression for the second order bulk moment:

$$
\begin{aligned}
r_{\mu_2^b} \approx r_{\mu_2} &= \sum_{k=1}^{N_d} \sum_{j=1}^{2} 2f_{k,j} k_{d_{k,j}} [I]_{k,j} \varphi_j + 2 \sum_{j=1}^{2} k_{p_j} [\lambda_{1,j}] [M]_j \varphi_j \\
&\quad + \sum_{j=1}^{2} k_{\mathrm{tc}_j} [\lambda_{1,j}]^2 \varphi_j + \sum_{j=1}^{2} \left( k_{p_j} + k_{\mathrm{fm}_j} \right) [\lambda_{0,j}] [M]_j \varphi_j
\end{aligned}
\tag{17}
$$

Note that in Eq. (17) $\mu_2$ is independent of the higher order moments (i.e., $\mu_3$).

## 2.3 Dynamic Molar Species and Energy Balances

Based on the postulated kinetic mechanism and the derived moment rate functions, the differential equations describing the time evolution of initiator(s), inhibitor, total monomer conversion, and live and dead moments in the batch reactor are written as:

$$\frac{d(I_i)}{dt} = -\sum_{j=1}^{2} k_{d_{i,j}} I_{i,j}; \quad j = 1, 2, \quad i = 1, 2, \ldots, N_d \tag{18}$$

$$\frac{d(Z)}{dt} = -\frac{1}{2} \sum_{j=1}^{2} k_{Z_j} Z_j \left[\lambda_{0,j}\right] \tag{19}$$

$$\frac{d(X)}{dt} = \sum_{j=1}^{2} k_{p_j} \frac{M_j}{M_0} \left[\lambda_{0,j}\right] \tag{20}$$

**Live, dead and bulk moments:**

$$\frac{d\left(\lambda_{i,j}\right)}{dt} = r_{\lambda_{i,j}}, (j = 1, 2) \tag{21}$$

$$\frac{d(\mu_i)}{dt} = r_{\mu_i} \tag{22}$$

$$\frac{d\left(\mu_i^b\right)}{dt} = r_{\mu_i^b} \tag{23}$$

Note that in the above design equations, the total number of kilomoles of species $i$ ($C_i$) is equal to the product term $V_j [C_i]$, where $V_j$ and $[C_i]$ are the volume of polymerization phase $j$ and concentration of species $i$, respectively. Based on the general kinetic scheme considered in this study, Eqs. (1)–(10), one can identify three additional structural characteristics of the polymer chains related to the number of LCBs, SCBs, and TDBs. To calculate the time variation of LCBs, SCBs, and TDBs per polymer molecule (i.e., $L_n$, $S_n$, and $T_n$, respectively) the following differential equations are used:

$$\frac{d([\text{LCB}])}{dt} = \frac{d(L_n[\mu_0])}{dt} = k_{fp_2} [\lambda_{0,2}][\mu_1] \tag{24}$$

$$\frac{d([\text{SCB}])}{dt} = \frac{d(S_n[\mu_0])}{dt} = \sum_{j=1}^{2} k_{b_j} [\lambda_{0,j}] \tag{25}$$

$$\frac{d([\text{TDB}])}{dt} = \frac{d(T_n[\mu_0])}{dt} = \sum_{j=1}^{2} \left( k_{fm_j} [M]_j [\lambda_{0,j}] + k_{td_j} [\lambda_{0,j}]^2 \right) \tag{26}$$

In Eqs. (11)–(26) the subscript $j$ refers to the polymerization phase, with $j = 1$ for the monomer-rich phase and $j = 2$ for the polymer-rich phase.

## 2.4  Calculation of Molecular Properties

The average molecular weights $M_n$ and $M_w$ can be expressed in terms of the moments of the TNCLDs of live and dead polymer chains as follows:

$$M_n = MW_m \frac{(\mu_1 + \lambda_{1,1} + \lambda_{1,2})}{(\mu_0 + \lambda_{0,1} + \lambda_{0,2})} = MW_m \frac{\mu_1}{\mu_0} \tag{27}$$

$$M_w = MW_m \frac{(\mu_2 + \lambda_{2,1} + \lambda_{2,2})}{(\mu_1 + \lambda_{1,1} + \lambda_{1,2})} = MW_m \frac{\mu_2}{\mu_1} \tag{28}$$

where $MW_m$ is the monomer molecular weight. The polydispersity index, a measure of the breadth of the molecular weight distribution, is given by the ratio $M_w/M_n$. The number density of LCBs, $L_d$, and SCBs, $S_d$, per 1,000 monomer units can be calculated from the following equations:

$$L_d = 1,000 \frac{L_n}{(\mu_1/\mu_0)}; \quad S_d = 1,000 \frac{S_n}{(\mu_1/\mu_0)} \tag{29}$$

## 2.5  Dynamic Energy Balances

The batch suspension polymerization reactor consists of a well-mixed jacketed vessel. Stirring is provided by a flat blade turbine, aided by four removable blade baffles. It is assumed that the reaction mixture is perfectly mixed. The reaction temperature is controlled by a cascade system of controllers that manipulate the flows of two streams (i.e., coolant and heating) entering the reactor jacket. Accordingly, one can derive the following dynamic energy balances for the reaction mixture, reactor metal wall, and fluid in the jacket [23]:

$$V_{mix}\rho_{mix}C_{p_{mix}} \frac{dT}{dt} = (-\Delta H_r)M_0 \frac{dX}{dt} - h_i A_i(T - T_m) + F_w c_{pw}\rho_w(T_0 - T)$$
$$- U_t A_t(T - T_a) \tag{30}$$

where $F_w$ represents the mass flow rate of the water added to the reaction mixture during polymerization to maintain a constant liquid level. $T_0$ is the inlet temperature of the water feed stream during polymerization. The term $U_t A_t(T - T_a)$ refers to heat losses from the reactor top. The meanings of all other symbols are given at the end of the chapter.

The reactor metal wall is treated as a lumped system:

$$V_m\rho_m C_{p_m}\frac{dT_m}{dt} = h_i A_i(T - T_m) - \frac{1}{4}A_0\sum_{i=1}^{4}h_{o,i}(T_m - T_{j,i}) \tag{31}$$

The total jacket volume is divided into four zones of equal volume:

$$V_j\rho_{w,i}C_{\text{pw},i}\frac{1}{4}\frac{dT_{j,i}}{dt} = \frac{1}{4}h_{o,i}A_0(T_m - T_{j,i}) + \frac{1}{4}U_a A_a(T_a - T_{j,i}) \\ +F_{w,j}C_{\text{pw},i}\rho_{w,i}(T_{j,i-1} - T_{j,i}) \tag{32}$$

Most correlations used for calculation of the inside film heat transfer coefficient in agitated vessels are of the following general form:

$$\text{Nu}_i = \frac{h_i D_R}{k_{\text{mix}}} = F(\text{Re}_i)^a(\text{Pr}_i)^{1/3}\left(\frac{\mu_{\text{mix}}}{\mu_{\text{mix},w}}\right)^b \tag{33}$$

The values of the parameters $F$, $a$, and $b$ can be found in heat-transfer textbooks for different types of agitators. For a flat-blade turbine, the recommended values for these parameters are $F = 0.54$, $a = 2/3$, and $b = 0.14$. The Reynolds and Prandtl numbers appearing in Eq. (33) are defined as:

$$\text{Re}_i = \frac{D_{\text{imp}}^2 N\rho_{\text{mix}}}{\mu_{\text{mix}}}; \quad \text{Pr}_i = \frac{C_{p_{\text{mix}}}\mu_{\text{mix}}}{k_{\text{mix}}} \tag{34}$$

where $D_{\text{imp}}$ and $N$ denote the impeller diameter and revolution number, respectively.

The outside film jacket heat transfer coefficient can be calculated from the following equation, assuming turbulent flow conditions for the coolant:

$$\text{Nu}_o = \frac{h_o D_{\text{eq}}}{k_w} = 0.023(\text{Re}_o)^{0.8}(\text{Pr}_o)^{1/3}\left(\frac{\mu_o}{\mu_w}\right)^{0.14} \tag{35}$$

where the dimensional numbers used are defined as:

$$\text{Pr}_o = \frac{C_{p_w}\mu_w}{k_w}; \quad \text{Re}_o = \frac{D_{\text{eq}}\rho_w u_w}{\mu_w} \tag{36}$$

$D_{\text{eq}}$ and $L_{\text{eq}}$ are the equivalent diameter and length, calculated in terms of the geometric characteristics of the reactor jacket.

A cascade control system consisting of a master proportional integral derivative controller and two slave proportional integral controllers is usually employed to maintain the polymerization temperature within $\pm0.1^\circ$C of the set-point value by manipulating the cold and hot water flow rates to the reactor jacket. The master controller monitors the reaction temperature and its output drives the set-point of the slave controller. The latter measures the outlet temperature of the coolant/

heating fluid in the jacket and drives the two separate control valves for the cold and hot water streams [23].

## 2.6 Phase Equilibrium Calculations

The vapor phase that occupies the free space on top of the liquid mixture in the reactor consists mainly of VCM, water vapor, residual air, and inert gases formed during the reaction. When a separate liquid monomer phase exists in the polymer-izing monomer droplets (i.e., in the conversion range $0 < X < X_f$), the reactor pressure is equal to the sum of the monomer and water partial pressures. It should be noted that a small amount of residual air might be present in the overhead vapor phase. However, because of the very low vacuum (e.g., less than 0.1 bar) typically applied to an industrial reactor before its loading, the amount of air in the overall overhead vapor mixture can be assumed negligible in the reactor pressure calcula-tion. In stage II of VCM polymerization, the polymer-rich phase remains saturated with monomer, reflecting the thermodynamic equilibrium between the two phases (i.e., monomer- and polymer-rich phases). However, when the separate monomer phase disappears (i.e., at the critical monomer conversion, $X = X_f$), the reactor pressure starts decreasing as a result of monomer transfer from the overhead vapor phase to the dispersed monomer-swollen polymer particles.

During VCM polymerization, the four phases are assumed to be in thermody-namic equilibrium. As a result, the fugacities of VCM in the four phases are equal:

$$\widehat{f}_m^{\ g} = \widehat{f}_m^{\ w} = \widehat{f}_m^{\ m} = \widehat{f}_m^{\ p} \tag{37}$$

Following the original developments of Xie et al. [39] and Kiparissides et al. [23], the fugacity coefficient of VCM, $\widehat{\varphi}_m$, in the gas phase is given by:

$$\ln\left(\widehat{\varphi}_m\right) = \ln\left(\frac{\widehat{f}_m^{\ g}}{P_m}\right) = \ln\left(\frac{\widehat{f}_m^{\ g}}{y_m P}\right) = \frac{P}{RT}\left[B_m + (1 - y_m)^2 \delta_{mw}\right] \tag{38}$$

where,

$$\delta_{mw} = 2B_{mw} - B_m - B_w \tag{39}$$

$P$ and $B_i$ are the reactor pressure and the second virial coefficient of the $i$ component, respectively. Assuming that the water vapor partial pressure is equal to its respective saturation value, the mole fraction of VCM in the vapor phase, $y_m$, can be calculated from the following equation:

$$y_m = (1 - y_w) = 1 - P_w^{\text{sat}}/P \tag{40}$$

The monomer activity, $a_m$, in the polymer-rich phase is given by the ratio of the monomer fugacity coefficient $(\hat{\varphi}_m)$ in the polymer phase to its corresponding value in the standard state. The latter is assumed equal to the fugacity of the pure monomer at the reaction temperature and respective monomer saturation pressure. According to the Flory–Huggins equation, monomer activity can be expressed in terms of the polymer volume fraction in the polymer-rich phase, $\varphi_2$, and the Flory–Huggins interaction parameter, $\chi$, [23]:

$$\ln(a_m) = \ln\left(\hat{f}_m^{p}/f_m^{0}\right) = \ln(1 - \varphi_2) + \varphi_2 + \chi\varphi_2^{2} \tag{41}$$

From Eqs. (37), (38), and (41), one can easily obtain the following equation for calculation of the total reactor pressure, $P$:

$$f_m^{0}\exp\left(\ln(1 - \varphi_2) + \varphi_2 + \chi\,\varphi_2^{2}\right) = y_m P\,\exp\left(\frac{P}{RT}\left[B_m + (1 - y_m)^2 \times \delta_{\text{mw}}\right]\right) \tag{42}$$

In the conversion range $0 < X < X_f$, the polymer phase is saturated with monomer. As a result, the monomer activity, $a_m$, is equal to one. Thus, from Eq. (41) for $a_m = 1$, the critical value of $\varphi_{2,C}$ can be obtained:

$$0 = \left(\ln(1 - \varphi_{2,C}) + \varphi_{2,C} + \chi\varphi_{2,C}^{2}\right) \tag{43}$$

### 2.6.1 Calculation of Monomer Distribution

To calculate the polymerization rates in the monomer- and polymer-rich phases, the VCM distribution in the four phases must be known. Assuming that the four phases in the system are in thermodynamic equilibrium, one can easily derive the following pseudo-steady-state monomer mass balance:

$$M_0(1 - X) = M_m + M_p + M_a + M_g \tag{44}$$

where the symbols $M_0$, $M_m$, $M_p$, $M_a$, and $M_g$ denote the total mass of VCM loaded in the reactor and the mass of VCM in the monomer ($m$), polymer ($p$), aqueous ($a$) and gas ($g$) phases, respectively.

During the first stage ($0 < X < 0.001$), no separate polymer phase exists (i.e., $M_p = 0$). Therefore, the mass of VCM in the monomer-rich phase is:

$$M_m = M_0(1 - X) - M_a - M_g \qquad (45)$$

According to Kiparissides et al. [23], the mass of VCM in the aqueous phase is given by:

$$M_a = K(P/P_m{}^{\text{sat}})\, W_{\text{wa}} \qquad (46)$$

where $K$ is the VCM solubility in the aqueous phase (i.e., $K = 0.0088$ g VCM/g $H_2O$), $P_m{}^{\text{sat}}$ is the saturated monomer vapor pressure at the polymerization temperature, and $W_{\text{wa}}$ is the mass of $H_2O$ in the aqueous phase. The latter is given by the following equation:

$$W_{\text{wa}} = W_w - \left(\widehat{f}_w{}^{g} \text{MW}_w V_g / RT\right) \qquad (47)$$

where $W_w$ is the total mass of water loaded in the reactor and $V_g$ the volume of the gas phase, calculated by the following equation:

$$V_g = \begin{aligned}&\left(V_R - (M_0/\rho_m) - (W_w/\rho_w) + M_0 X\left(\rho_m{}^{-1} - \rho_p{}^{-1}\right)\right)\\&\times \Big/\left(1 - \left(\widehat{f}_m{}^{g}\text{MW}_m/\rho_m + \widehat{f}_w{}^{g}\,\text{MW}_w/\rho_w\right)/RT\right)\end{aligned} \qquad (48)$$

$V_R$ is the total reactor volume and $\rho_m$, $\rho_p$, $\rho_w$ are the corresponding monomer, polymer, and water densities. $\text{MW}_m$ and $\text{MW}_w$ denote the molecular weights of monomer and water, respectively. The fugacity of water in the vapor phase, $\widehat{f}_m{}^{g}$, is equal to the total reactor pressure, $P$, minus the fugacity of the monomer in the gas phase, $\widehat{f}_m{}^{g}$:

$$\widehat{f}_m{}^{g} = P - \widehat{f}_m{}^{g} \qquad (49)$$

Accordingly, the mass of VCM in the gas phase can be calculated in terms of its fugacity, $\widehat{f}_m{}^{g}$, the volume of the gas phase, $V_g$, and the temperature, $T$:

$$M_g = \widehat{f}_m{}^{g}\, \text{MW}_m V_g / RT \qquad (50)$$

Following similar considerations, monomer distributions in the monomer-rich and polymer-rich phases during stage II of VCM polymerization (i.e., $0.001 < X < X_f$), are given by the following equations:

$$M_m = M_0\left(1 - \frac{X}{X_S}\right) - M_a - M_g \qquad (51)$$

$$M_p = M_0 \frac{X}{X_S}(1 - X_S) \tag{52}$$

$$X_S = \varphi_{2,C}\, \rho_P / \left(\varphi_{2,C}\, \rho_P + (1 - \varphi_{2,C})\rho_m\right) \tag{53}$$

The values of $M_a$ and $M_g$ are calculated from Eqs. (46) and (51), respectively. During this polymerization stage, monomer concentrations in the monomer-rich phase (indicated by subscript 1) and polymer-rich phase (indicated by subscript 2) are given by:

$$[M_m] = \rho_m / \mathrm{MW}_m; \quad [M_p] = M_p / (\mathrm{MW}_m V_2) \tag{54}$$

Accordingly, the volumes of each phase are:

$$V_1 = M_m / \rho_m; \quad V_2 = (M_p / \rho_m) + (M_0 X / \rho_P) \tag{55}$$

In stage III $(X > X_f)$, the separate monomer-rich phase disappears and polymerization takes place only in the polymer-rich phase. Accordingly, monomer distributions in the monomer, polymer, and gas phases are given by the following equations:

$$M_m = 0; \quad M_p = M_0(1 - X) - M_a - M_g \tag{56}$$

$$M_g = \left(\widehat{f}_m{}^g\, \mathrm{MW}_m / (\mathrm{RT})\right)\left(V_g(X_f) + M_0(X - X_f)\left(\frac{1}{\rho_m} - \frac{1}{\rho_p}\right)\right) \tag{57}$$

The monomer conversion, $X_f$, at which the monomer-rich phase disappears can be calculated from Eq. (44), by setting the value of $M_m$ equal to zero. Thus, the following expression is obtained:

$$\begin{aligned} &X_f \\ &= \frac{X_S\left(M_0 - \mathrm{KW}_{ww} - \left(\widehat{f}_m{}^g\, \mathrm{MW}_m\left(V_R - \frac{M_0}{\rho_m} - \frac{W_w}{\rho_w}\right)\right)\middle/\left(RT - \left(\frac{\widehat{f}_m{}^g\, \mathrm{MW}_m}{\rho_m} + \frac{\widehat{f}_m{}^g\, \mathrm{MW}_m}{\rho_w}\right)\right)\right)}{M_0\left(1 + \widehat{f}_m{}^g\, \mathrm{MW}_m\, X_S((1/\rho_m) - (1/\rho_p))\middle/\left(RT - \left(\left(\widehat{f}_m{}^g\, \mathrm{MW}_m/\rho_m\right) + \left(\widehat{f}_m{}^g\, \mathrm{MW}_m/\rho_w\right)\right)\right)\right)} \end{aligned} \tag{58}$$

## 2.7 Diffusion-Controlled Reactions

At high monomer conversions, almost all elementary reactions become diffusion-controlled. Specifically, the initiation, propagation, and termination reactions are related to the well-known phenomena of cage-, glass-, and gel-effects, respectively.

Diffusion-controlled reactions can be described quantitatively using the generalized free-volume theory [40–42].

In the present study, the simplified free-volume model of Xie et al. [7] was employed to describe diffusion-controlled reactions in the free-radical suspension polymerization of VCM. Accordingly, the diffusion-controlled termination rate constant in the polymer-rich phase, $k_{t2}$, is expressed as follows:

$$k_{t2} = k_{t20} \exp\left(-A\left(\frac{1}{V_f} - \frac{1}{V_f^*}\right)\right) \tag{59}$$

where $k_{t20}$ is the termination rate constant in the polymer-rich phase for $X < X_f$. Term $V_f$ is the free volume of the mixture in the polymer-rich phase, given by the following equation:

$$V_f = \varphi_2 V_{fp} + (1 - \varphi_2)V_{fm} \tag{60}$$

where $V_{fp}$ and $V_{fm}$ are the free volumes of the polymer and monomer, respectively, and $\varphi_2$ is the volume fraction of polymer in the polymer-rich phase. $V_f^*$ is the value of $V_f$ at the critical monomer conversion, $X_f$. Similarly, the diffusion-controlled propagation rate constant in the polymer-rich phase, $k_{p2}$, is written:

$$k_{p2} = k_{p1} \exp\left(-B\left(\frac{1}{V_f} - \frac{1}{V_f^*}\right)\right) \tag{61}$$

where $k_{p1}$ is the propagation rate constant in the monomer-rich phase. The variation in initiator efficiency with respect to monomer conversion can be calculated using the followng equation:

$$\left(k_p f^{1/2}\right)_X = \left(k_p f^{1/2}\right)_{X_f} \exp\left(-B_f\left(\frac{1}{V_f} - \frac{1}{V_f^*}\right)\right) \tag{62}$$

The temperature-dependant constants $A$, $B$, and $B_f$ in Eqs. (59), (61), and (62) can be estimated by fitting the model predictions to available experimental rate data [7, 23, 24].

## 2.8    Results and Discussion on Polymerization Kinetics

The predictive capabilities of the developed kinetic model were tested by direct comparison of model predictions to a comprehensive series of experimental measurements on polymerization rate, monomer conversion, reactor pressure, etc. The free-radical suspension polymerization of VCM was assumed to proceed under

**Table 1** Kinetic rate coefficients for the free-radical polymerization of VCM [23]

| |
|---|
| $k_{p1} = k_{p2} = 3 \times 10^9 \exp(-3,320/T) \ (\text{m}^3 \times \text{kmol}^{-1} \times \text{min}^{-1})$ |
| $k_{fm1} = k_{fm2} = 5.78 \ \exp(-2,768/T)k_{P1}(\text{m}^3 \times \text{kmol}^{-1} \times \text{min}^{-1})$ |
| $k_{b1} = k_{b2} = 0.014 \times k_{P1}(\text{min}^{-1})$ |
| $k_{tcj} = k_{tj}/2 \, ; \ \ k_{tdj} = k_{dj}/2 \, ; \ \ j = 1, 2$ |
| $k_{t1} = 2\frac{k_{p1}^2}{K_c} \ (\text{m}^3 \times \text{kmol}^{-1} \times \text{min}^{-1}) \, ;$ <br> $K_c = 6.08 \times 10^{-3} \ \exp\left(-5,740\left(\frac{1}{T} - \frac{1}{T_0}\right)\right)$ |
| $(k_{t1}/k_{t20})^{1/2} = 24 \exp(1,007(T^{-1} - T_0^{-1})), \ \ \ T_0 = 333.15 \ (\text{K})$ |
| $(-\Delta H_r) = 106 \ (\text{kJ kmol}^{-1})$ |

**Table 2** Physical and thermodynamic properties of water, VCM, and PVC [24]

| |
|---|
| $\rho_m = 947.1 - 1.746\theta - 3.24 \times 10^{-3}\theta^2 (\text{kg m}^{-3})$ <br> $\rho_w = 1011.0 - 0.4484\theta \ (\text{kg m}^{-3})$ <br> $\rho_P = 10^3 \exp(0.4296 - 3.274 \times 10^{-4}T)(\text{kg m}^{-3})$ |
| $P_w^{\text{sat}} = \exp(72.55 - 7,206.7/T - 7.1386 \times \ln(T) + 4.046 \times 10^{-6}T^2)(\text{Pa})$ |
| $P_m^{\text{sat}} = P_{Cm} \exp\left(\frac{1}{1-X_r}\left(-6.5008 \, X_r + 1.21422 \, X_r^{1.5} - 2.57867 \, X_r^3 - 2.00937 \, X_r^6\right)\right) \times 10^5 \ (\text{Pa})$ |
| $X_r = 1 - \frac{T}{T_{Cm}}$ |
| $T_{Cm} = 425 \, , \ \ T_{Cw} = 647.3 \ (\text{K})$ <br> $P_{Cm} = 51.5 \, , \ \ P_{Cw} = 221.2 \ (\text{bar})$ <br> $V_{Cm} = 169 \, , \ \ V_{Cw} = 57.1 \ (\text{cm}^3 \, \text{mol}^{-1})$ <br> $\omega_m = 0.122 \, , \ \ \omega_w = 0.344$ <br> $Z_{Cm} = 0.265 \, , \ \ Z_{Cw} = 0.235$ |

**Table 3** Parameters used in the diffusion model [7]

| |
|---|
| $V_{fm} = 0.025 + a_m(T - T_{gm})$ |
| $V_{fp} = 0.025 + a_p(T - T_{gp})$ |
| $a_m = 9.98 \times 10^{-4} \, ; \ \ a_p = 5.47 \times 10^{-4}(\text{K}^{-1})$ |
| $T_{gm} = 70 \ (\text{K}) \, ; \ \ T_{gp} = 87.1 - 0.132(T - 273.15) \ (^{\circ}\text{C})$ |
| $A = 6.64 \times 10^6 \exp(-5,080/T)$ |
| $B = 0.93 \times 10^5 \exp(-3,850/T)$ |
| $B_f = 4.74 \times 10^4 \exp(-5,064/T)$ |

isothermal conditions in the presence of monofunctional initiators such as di (2-ethylhexyl) peroxydicarbonate (PDEH), 3-hydroxy-1,1-dimethylbutyl peroxyneodecanoate (LUP610), and lauroyl peroxide (LP40).

Table 1 lists the numerical values of the kinetic rate coefficients employed in the computer simulations. Values for the rate coefficients for chain transfer to monomer, propagation, backbiting, and termination reactions were taken from the work of Kiparissides et al. [23]. Table 2 reports the physical and thermodynamic properties of the VCM/PVC/H$_2$O system. Parameters used in the free-volume model are given in Table 3.

In Figs. 5, 6, and 7, experimental and model results for monomer conversion, polymerization rate, and reactor pressure are plotted with respect to the

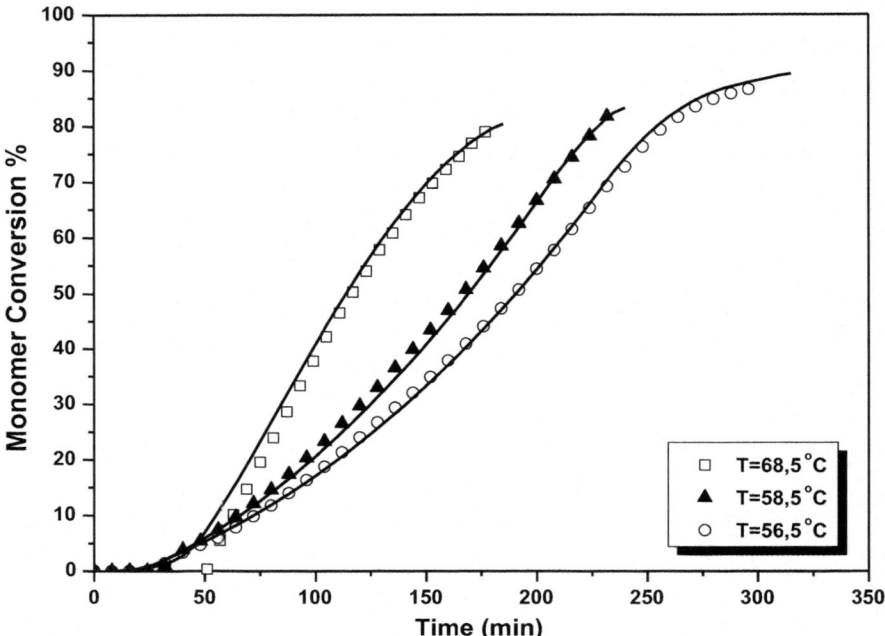

**Fig. 5** Predicted (*lines*) and experimental (*points*) conversion time histories for polymerization temperatures of 68.5, 58.5, and 56.5°C; $I_{0,PDEH} = 0.606$ g/kg VCM

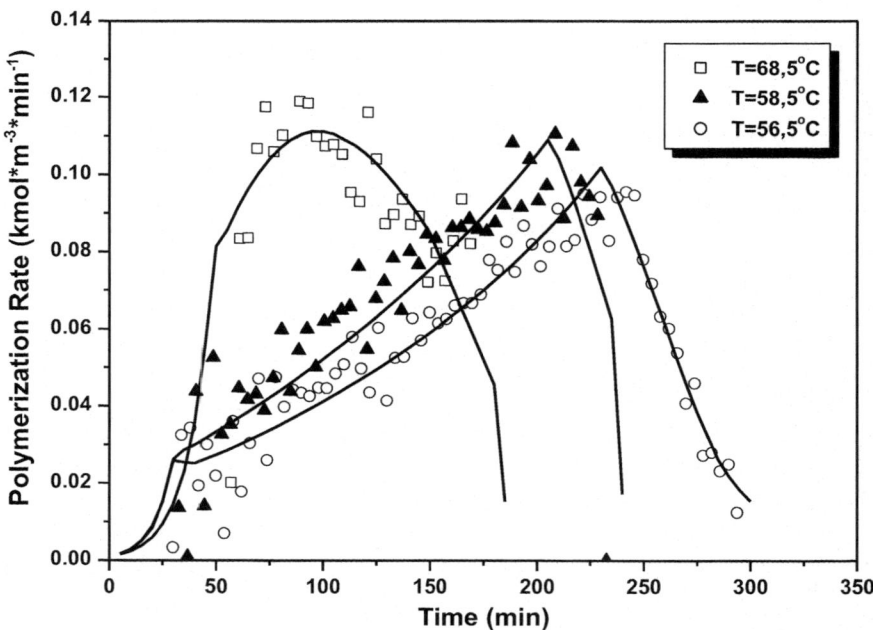

**Fig. 6** Predicted (*lines*) and experimental (*points*) polymerization rates for polymerization temperatures of 68.5, 58.5, and 56.5°C; $I_{0,PDEH} = 0.606$ g/kg VCM

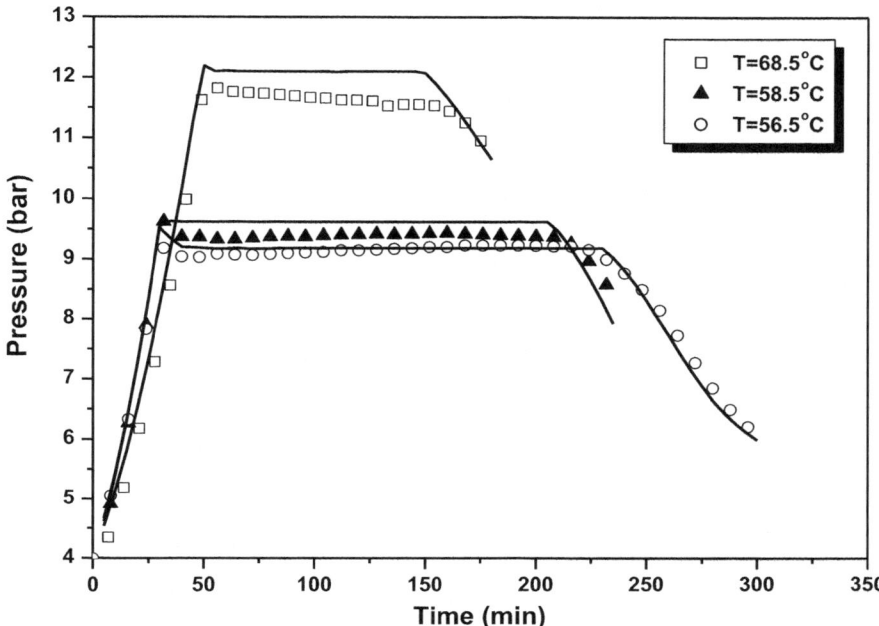

**Fig. 7** Predicted (*lines*) and experimental (*points*) reactor pressures with respect to polymerization time for polymerization temperatures of 68.5, 58.5, and 56.5°C; $I_{0,\text{PDEH}} = 0.606$ g/kg VCM

polymerization time at three different temperatures (68.5, 58.5, and 56.5°C). The discrete points represent experimental measurements and continuous lines denote the respective model predictions. The free-radical suspension polymerization of VCM was initiated via the thermal decomposition of PDEH. In the reported runs, the initiator concentration was 0.606 g/kg VCM and the mass ratio of water to VCM was 1.55. It is apparent that an excellent agreement exists between model predictions and experimental results. The observed small discrepancies in the polymerization rate and reactor pressure are primarily the result of measurement noise in the experimental measurements.

In Fig. 8, the effect of polymerization temperature on $M_n$ and $M_w$ of PVC is shown for the two initiators. The molecular weight averages decrease as the polymerization temperature increases. On the other hand, at the same polymerization temperature, the initiator type does not significantly affect the values of $M_n$ and $M_w$ because their values primarily depend on the transfer to monomer rate constant.

In Fig. 9, experimental data collected from an industrial-scale PVC batch suspension polymerization reactor (i.e., 125 m$^3$) and model predictions on monomer conversion and reactor temperature are plotted with respect to polymerization time. The free-radical suspension polymerization of VCM was initiated via thermal decomposition of PDEH. In the reported run, the initiator concentration was 0.646 g/kg VCM and the mass ratio of water to VCM was 0.32. Excellent

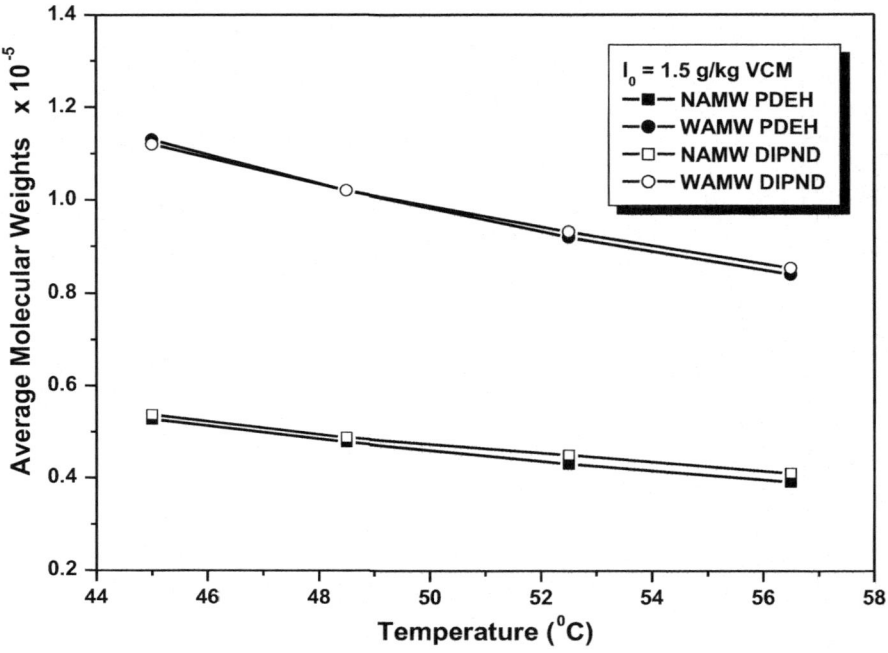

**Fig. 8** Predicted number- $(M_n)$ and weight-average $(M_w)$ molecular weights with respect to polymerization temperature. Comparison between DIPND and PDEH initiators; $I_{0,DIPND} = I_{0,PDEH} = 1.5$ g/kg, mass of water $= 1.55$ kg/kg VCM

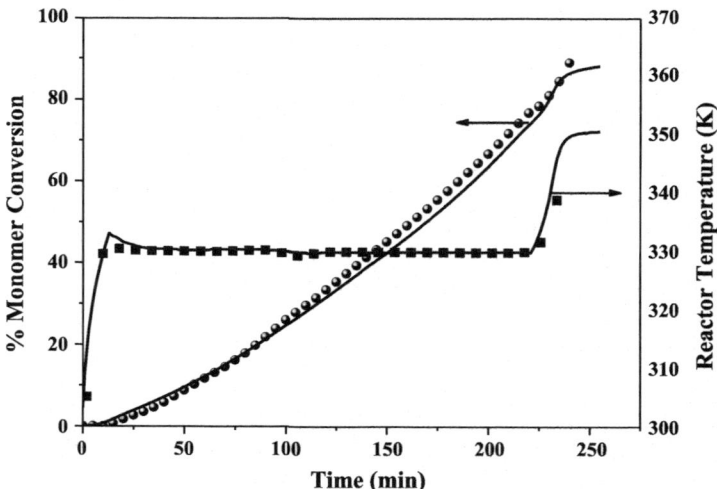

**Fig. 9** Predicted (*lines*) and experimental (*points*) monomer conversion and reactor temperature time histories; $M_{0,PDEH} = 0.646$ g/kg VCM, mass of water$= 0.32$ kg/kg VCM

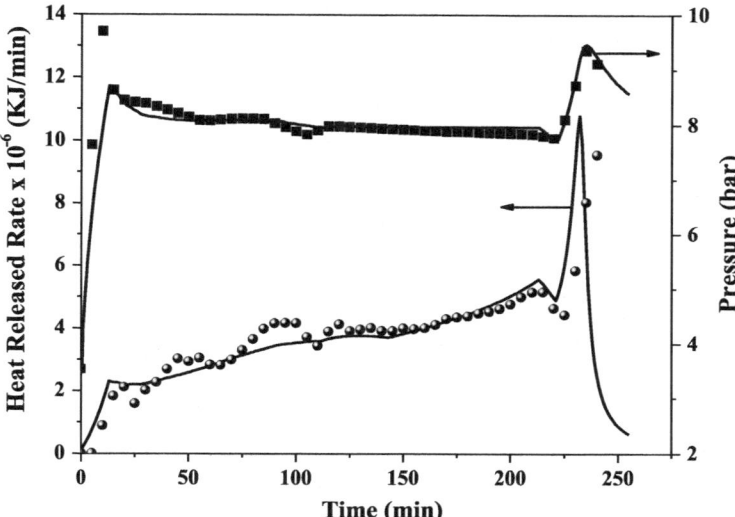

**Fig. 10** Predicted (*lines*) and experimental (*points*) heat released and pressure time histories (experimental conditions as for Fig. 9)

agreement exists between the model predictions and the experimental measurements. Note that at approximately 221 min after the polymerization started, the condenser operation was stopped and cooling water supply to the circulation loop was almost completely shut down. As a result, the temperature in the reactor increased because of self-heating of the reaction mixture due to the strongly exothermic reaction. This strategy was applied to increase the initiator consumption rate and, thus, monomer conversion during the final stages of polymerization. In Fig. 10, experimental and model predictions for reactor pressure and heat release rate are plotted as a function of polymerization time. As can be seen, the simulation results are in good agreement with experimental measurements. The heat release rate increases with the increase in polymerization rate until critical monomer conversion (i.e., $X_f \approx 70\%$). After critical conversion, the heat release rate and the reactor pressure decrease because of disappearance of the monomer-rich phase. This is followed by monomer transfer from the gas phase to the polymer-rich phase. The decrease in total reactor pressure stops when the polymerization temperature starts increasing rapidly as a result of adiabatic reactor operation (Fig. 9). However, as the monomer mass in the gas phase becomes depleted, the pressure again decreases together with the heat release rate until the batch operation terminates.

## 3 Dynamic Evolution of Particle Size Distribution in Suspension PVC

The "powder" suspension polymerization process is the most important polymerization process for PVC manufacture. The main advantage of this process is that large (e.g., 300–500 μm) porous polymer particles can be produced. The process exhibitsg a fast residual monomer removal rate and a large plasticizer uptake capacity. The production of polymer particles with desired PSD and porosity can be achieved by changing the type and concentration of stabilizer(s) in the polymerization recipe as well as the agitation rate, without affecting the molecular properties of the product. Polymerization is commonly carried out isothermally, at temperatures of 45–70°C, depending on the desired molecular weight of PVC. VCM is an extremely volatile compound and its vapor pressure varies approximately from 8 to 12 bar in that temperature range [43].

In free-radical VCM polymerization, the first polymer chains produced inside the monomer droplets precipitate out to form unstable polymer microdomains with diameters of 10–20 nm (see Fig. 11). These microdomains exhibit limited stability and, thus, aggregate to form the nuclei of primary particles, also called domains (at monomer conversions between 0.01 and 10%). The initial size of these domains is 80–100 nm. The growth of these domains via polymerization of absorbed monomer or/and by aggregation with other domains results in primary particles with diameters of 100–200 nm. At critical monomer conversion (10–30%), massive aggregation of the primary particles leads to the formation of a three-dimensional (3D) polymer skeleton. The primary particles continue to grow until disappearance of the free monomer phase (i.e., at a fractional monomer conversion, $X_f$). In the presence of secondary stabilizers at low agitation speeds, the aggregation and subsequent fusion of primary particles can be limited, although individual primary particles can continue to grow up to 1–1.5 μm in diameter. In the latter case, primary particles can pack closely together without any significant aggregation. This results in a close-packed structure of low porosity. The porosity of PVC grains increases as the agitation rate increases. Strong agitation can also favor aggregation of individual polymerizing monomer droplets, leading to the formation of irregular PVC grains with diameters of 50–250 μm. The main difference between the bulk and the suspension process is that agitation is used to control not only aggregation of the primary particles but also the size distribution of the final grains [43].

As a result of the above mechanism, the polymerizing VCM droplets lose their viscous characteristics at relatively low monomer conversions, whereas at larger monomer conversions (i.e., $X > 30\%$) they behave like rigid spheres because of the presence of an internal continuous polymer skeleton. Above a critical monomer conversion ($X_c \sim 30\%$), the volume contraction of the polymerizing particles stops, which partially explains the appearance of internal particle porosity. Note that the polymer density is approximately 40% higher than the monomer density.

In VCM suspension polymerization, two types of stabilizers (i.e., primary and secondary) are used [44]. The main function of the primary surface-active agents is

**Fig. 11** Evolution of primary PVC particles

Coiled Macroradicals

Micro-Domains
(x<1%)

Domains
(x=1-2%)

Primary Particles

Primary
Particle
Agglomerates
x=4-10%

Fused
Agglomerates
x=90%

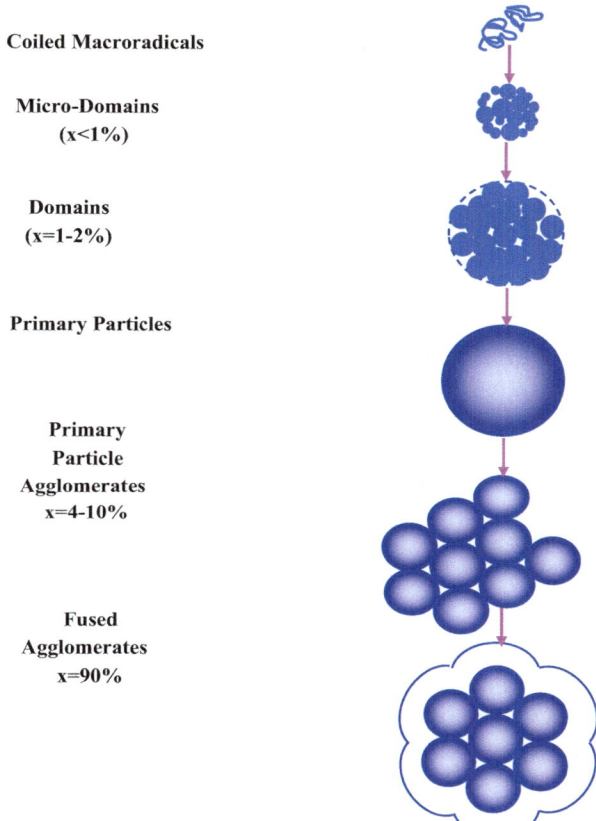

to control the grain size, but they also affect the internal grain porosity. Secondary stabilizers are surface-active agents with a higher lipophilic content (e.g., PVA stabilizers with low degree of hydrolysis and cellulose ethers with high degree of substitution of the hydroxyl groups). They are mainly soluble in the VCM droplets and adsorb onto the surface of PVC primary particles. The increased stability of the primary particles, imparted by the adsorbed secondary stabilizers, results in a decrease in the aggregation rate of primary particles with a concomitant the decrease in PVC grain porosity.

One of the most important issues in suspension polymerization is control of the final PSD [25, 45–47]. The initial monomer droplet size distribution (DSD) and the final polymer PSD depend on the type and concentration of the surface-active agents, the quality of agitation, and the physical properties (e.g., density, viscosity, and interfacial tension) of the continuous and dispersed phases. The dynamic evolution of the droplet/particle size distribution is controlled by two dynamic processes, namely, the rates of droplet/particle breakage and coalescence. The former mainly occurs in regions of high shear stress (i.e., near the agitator blades) or as a result of turbulent velocity and pressure fluctuations along the surface of a drop. The latter is either increased or decreased by the turbulent flow field and can be assumed negligible for very dilute dispersions at sufficiently high concentrations of surface-active agents [48].

## 3.1 Surface-Active Agents

Surface-active agents play a very important role in the stabilization of liquid–liquid dispersions. They can be water-soluble copolymers such as poly(vinyl alcohol) (PVA) and cellulose ethers or colloidal inorganic powders (Pickering dispersants such as tricalcium phosphate, barium sulfate, and calcium carbonate). The former mainly consist of hydrophilic and hydrophobic (lipophilic) monomer units. The lipophilic segments of polymeric stabilizers are soluble in the organic monomer phase, whereas the hydrophilic units tend to remain in the continuous aqueous phase. These stabilizers reduce the drop/particle coalescence rate because of steric repulsive forces [49]. The presence of a protective colloidal film around the droplets prolongs their contact time before coalescence, thus increasing the probability of separation of drops by agitation. Hartland [50] argued that stabilizers increase the interfacial viscosity and lower the interfacial tension. The most important factor in determining the effectiveness of a polymeric stabilizer is its hydrophilic–lipophilic balance, whereas the molecular mass of the stabilizer is less significant [51].

One of the most commonly used stabilizers in suspension polymerization is poly (vinyl acetate) that has been partially hydrolyzed to PVA. By varying the acetate content (i.e., degree of hydrolysis; DH), it is possible to alter the hydrophobicity of PVA and, thus, the conformation and surface activity of the polymer chains at the monomer–water interface [52]. The solubility of PVA in water depends on the overall degree of polymerization (i.e., molecular weight of PVA), the sequence chain length distribution of the vinyl alcohol and vinyl acetate units in the copolymer, DH, and temperature. Depending on the agitation rate and the concentration and type of surface-active agent, the average droplet size can exhibit a U-shape variation with respect to agitation speed (see Figs. 12 and 13). This U-type behavior has been confirmed both experimentally and theoretically and is attributed to the balance between breakage and coalescence rates for monomer drops. The initial decrease in drop size with increasing impeller speed can be explained by the higher drop breakage frequency due to the higher shear stresses. However, as the interfacial area increases because of formation of a larger number of small droplets in the system, the stabilizer concentration cannot fully cover the larger surface area of the smaller droplets. As a result, the droplet coalescence rate increases, leading to a final increase in the drop/particle size.

Water-soluble substituted celluloses are also used as stabilizers in PVC suspension polymerization. These stabilizers are soluble in both vinyl chloride and the aqueous phase [53]. Consequently, the stabilizer can affect the stability of the primary particles inside the polymerizing monomer droplets and, thus, the final

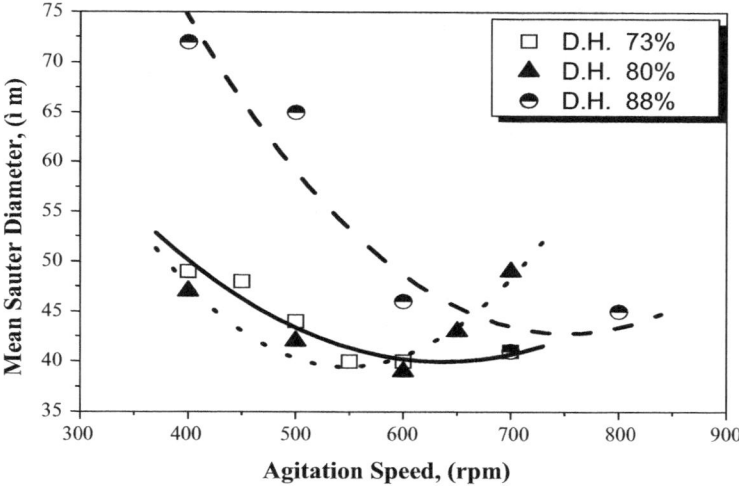

**Fig. 12** Effect of degree of hydrolysis of PVA and agitation rate on the steady-state Sauter mean droplet diameter in a liquid–liquid dispersion for three PVA grades

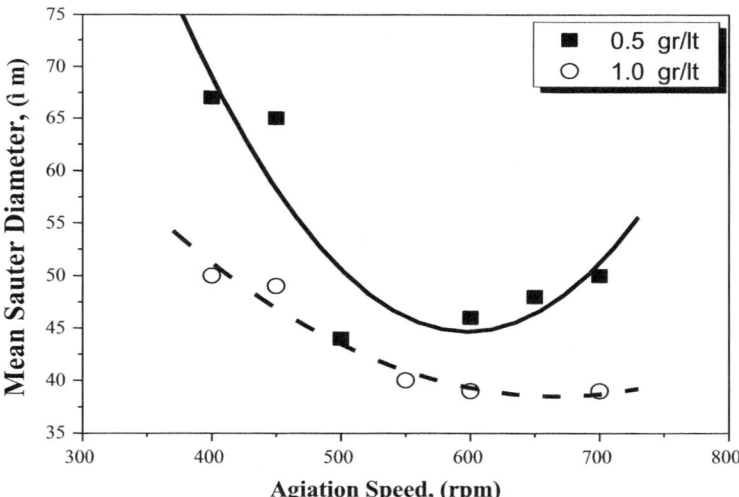

**Fig. 13** Effect of PVA concentration and agitation rate on the steady-state Sauter mean droplet diameter in a liquid–liquid dispersion for three PVA grades

porosity of the PVC grains. Hydroxypropyl methylcellulose (HPMC) is a cellulose ether, produced by reacting cellulose with propylene oxide and methyl chlorine in an alkaline medium. As a result, a fraction of the hydroxyl groups (hydrophilic groups) of the cellulose are substituted by hydroxypropyl and methyl groups (lipophilic groups). Cellulose ethers are generally characterized by their solution viscosity, chemical nature of the substituent, degree of substitution, purity, rheo-logical properties, solubility in the aqueous phase, and compatibility with the polymer. It has been reported that HPMC/VCM compatibility depends on the molecular weight and the degree of substitution of the cellulose derivative. Thus, HPMC of low molecular weight has a higher compatibility with VCM and results in a higher porosity of the final PVC resin [54].

Pickering stabilizers are inorganic solids that are insoluble in the aqueous phase. Their main advantage is that they can be removed easily from the final particulate product (e.g., by dilute acid), which improves the clarity and transparency of the polymer. In addition, the amount of polymer deposited on the inside reactor wall and other parts decreases considerably, which improves the heat transfer rate from the reaction medium to the coolant in the jacket. Finally, it should be mentioned that inorganic powders are usually relatively cheap [45].

The sorption kinetics of stabilizer molecules from the continuous phase to the organic–water interface changes with time. The time required for the sorption process to reach steady state is controlled by the transfer of stabilizer molecules from the continuous phase to the droplet surface and their subsequent reconformation and rearrangement at the organic–water interface. As the stabilizer concentration increases, the time required for the system to reach equilibrium is reduced, indicating an increased polymer diffusion rate [55]. Nilsson et al. [56] argued that the stabilizer molecules diffuse quickly to the liquid–liquid interface, but not in the most thermodynamically stable conformation. Thus, rearrangement of the stabilizing molecules takes place until the system (interface) reaches thermo-dynamic equilibrium.

Chatzi and Kiparissides [52] studied the dispersion of $n$-butyl chloride in water in the presence of various PVA stabilizers and observed two critical PVA concen-trations at which the interfacial tension exhibited a sharp change. They found that, at low PVA concentrations (e.g., less than 0.001 g/L), the interfacial tension was relatively independent of PVA concentration for all types of PVA studied. At higher concentrations, the interfacial tension decreased almost linearly with the PVA concentration on a semi-log scale. This convex behavior was also observed by Lankveld and Lyklema [57] for a paraffin-oil/water system in the presence of a PVA stabilizer with DH of 88%. In Fig. 14, the measured interfacial tension of an $n$-butyl/water system is plotted as a function of PVA concentration [52]. Assuming a linear dependence of the interfacial tension, $\sigma$, with respect to surface coverage, $\vartheta$, the experimental results can be fitted to an ideal Langmuir-like adsorption isotherm in terms of , $\vartheta$ and the PVA concentration:

$$\sigma = 35 - 32\vartheta; \quad \vartheta = \text{PVA}/(0.001 + \text{PVA}) \tag{63}$$

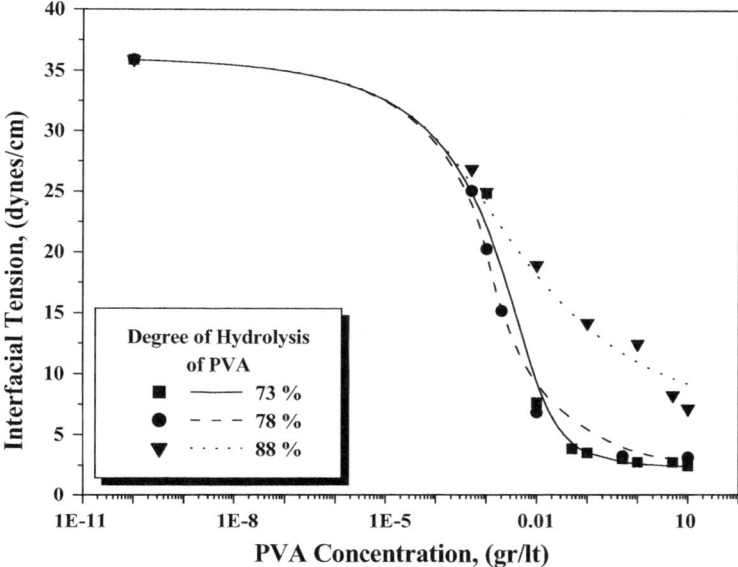

**Fig. 14** Variation in the equilibrium interfacial tension of *n*-butyl chloride/water with respect to the PVA concentration at 20°C for three different PVA grades

## 3.2 Monomer Droplet/Polymer Particle Population Balance Model

With regard to the droplet/particle breakage and coalescence, the suspension polymerization process can be divided into three stages [51, 54, 58]. During the initial low-conversion (i.e., low viscosity) stage, drop breakage is the dominant mechanism. As a result, the initial DSD shifts to smaller sizes. During the second or sticky stage of polymerization, the drop breakage rate decreases and drop/particle coalescence becomes the dominant mechanism. Thus, the average particle size starts to increase. In the third stage, the PSD reaches its identification point, whereas the polymer particle size decreases slightly because of shrinkage (i.e., the polymer density is greater than the monomer density). In VCM powder polymerization, at monomer conversions around 10–30%, a continuous polymer network forms inside the polymerizing monomer droplets that significantly reduces the drop/particle coalescence rate [59]. Cebollada et al. [54] reported that the PSD is essentially established up to monomer conversions of about 35–40% (i.e., end of the second stage).

A generalized population balance model is proposed to describe the dynamic evolution of PSD in batch suspension polymerization reactors. The model takes into account dynamic evolution of the physical and transport properties of the continuous and disperse phases, monomer conversion, turbulent intensity of the flow field, and their relative effects on the rates of breakage and coalescence of droplets/particles.

To follow the dynamic evolution of PSD in a particulate process, a population balance approach is commonly employed. Distribution of the droplets/particles is considered to be continuous in the volume domain and is usually described by a number density function, $n(v, t)$. Thus, $n(v, t)dv$ represents the number of particles per unit volume in the differential volume size range $(v, v+dv)$. For a dynamic particulate system, undergoing simultaneous particle breakage and coalescence, the rate of change of the number density function with respect to time and volume is given by the following nonlinear integro-differential PBE [25, 47, 58, 60]:

$$\frac{d[n(v,t)]}{dt} = \int_v^{v_{max}} \beta(u,v)\, u(u)\, g(u)\, n(u,t)du + \int_{v_{min}}^{v/2} k(v-u,u)\, n(v-u,t)\, n(u,t)du$$

$$-n(v,t)\, g(v) - n(v,t) \int_{v_{min}}^{v_{max}} k(v,u)\, n(u,t)du$$

$$(64)$$

The first term on the right-hand side of Eq. (64) represents the generation of droplets in the size range $(v, v+dv)$ as a result of drop breakage. $\beta(u, v)$ is a daughter drop breakage function, accounting for the probability that a drop of volume $v$ is formed via the breakage of a drop of volume $u$. The function $u(u)$ denotes the number of droplets formed by breakage of a drop of volume $u$ and $g(u)$ is the breakage rate of drops of volume $u$. The second term on the right-hand side of Eq. (64) represents the rate of generation of drops in the size range $(v, v+dv)$ as a result of drop coalescence. $k(v, u)$ is the coalescence rate between two drops of volume $v$ and $u$. The third and fourth terms represent the drop disappearance rates as a result of drop breakage and coalescence, respectively. Equation (64) satisfies the following initial condition at $t = 0$:

$$n(v,0) = n_0(v) \qquad (65)$$

where $n_0(v)$ is the initial drop size distribution of the disperse phase. The initial monomer drop size distribution is generally considered to follow a normal distribution around a mean value $V_0$ and standard deviation $\sigma_0$ [25].

## 3.3    Monomer Droplet Breakage Process

It has been postulated that droplet breakage in turbulent flow fields is caused by viscous shear forces, turbulent pressure fluctuations [61, 62], or/and relative velocity fluctuations [63]. When drop breakage occurs by viscous shear forces, the monomer droplet is first elongated into two fluid lumps separated by a liquid thread. Subsequently, the deformed monomer droplet breaks into two almost equally sized drops, corresponding to the fluid lumps, and a series of smaller droplets

**Table 4** Droplet size correlations for liquid–liquid dispersions in stirred tanks

| Researchers | Correlation | Operating conditions | |
|---|---|---|---|
| | | $\varphi$ (volume fraction of dispersed-phase) | N (impeller speed) (rpm) |
| Chen and Middleman [65] | $d_{32} = 0.053 N_{We}^{-0.6}$ | | |
| Doulah [66] | $d_{32} = C(1 + 3\phi) N_{We}^{-0.6}$ | | |
| Coulaloglou and Tavlarides [67] | $d_{32} = 0.081(1 + 4.47\phi) N_{We}^{-0.6}$ | 0.025–0.15 | 190–310 |
| Wang and Calabrese [68] | $d_{32}/D_I = 0.053(1 + 0.97 N_{Vi}^{0.79})^{0.6} N_{We}^{-0.6}$ | <0.002 | 80–280 |
| Calabrese et al. [69] | $d_{32}/D_I = 0.053(1 + 0.91 N_{Vi}^{0.84})^{0.6} N_{We}^{-0.6}$ $N_{Vi} = (\rho_c/\rho_d)^{1/2}(\mu_d \varepsilon^{1/3} d_{max}^{1/3}/\sigma)$ | <0.2 | 80–280 |
| Lagisetty et al. [70] | $d_{32}/D_I = 0.083(1 + 4.0\phi)^{1.2} N_{We}^{-0.6}$ | | |
| Laso et al. [71] | $d_{32}/D_I = 0.118 \phi^{0.27}(\mu_d/\mu_c)^{-0.056} N_{We}^{-0.4}$ | 0.10 | 370–540 |
| Chatzi et al. [72] | $d_{32}/D_I = 0.0165(1 + 11.94\phi) N_{We}^{-0.4}$ (coalescence dominant) $d_{32}/D_I = 0.056(1 + 10.97\phi) N_{We}^{-0.4}$ (breakage dominant) | 0.01–0.03 | 150–300 |
| Chatzi et al. [73] | $d_{32}/D_I = 0.045 N_{We}^{-0.4}$ | 0.01 | 200–300 |
| Zerfa and Brooks [74] | $d_{32}/D_I = 0.027(1 + 3.1\phi) N_{We}^{-0.6}$ | 0.01–0.4 | 250–800 |

corresponding to the liquid thread. This is known as "thorough breakage." On the other hand, a droplet suspended in a turbulent flow field is exposed to local pressure and relative velocity fluctuations. For nearly equal densities and viscosities of the two liquid phases, the droplet surface can start oscillating. When the relative velocity is close to that required to make a drop marginally unstable, a number of small droplets are stripped from the initial one. This situation of drop breakage is referred to "erosive breakage." Erosive drop breakage is the dominant mechanism of low-coalescence systems that exhibit a characteristic bimodality in the PSD [48, 64].

The first approaches to modeling the drop breakage process in liquid–liquid dispersions were based on the Weber number ($N_{We}$) for calculation of the mean drop diameter (see Table 4), as well as on a maximum stable drop diameter for breakage to occur and a minimum drop diameter above which coalescence takes place [69, 70, 75, 76]. Both maximum and minimum drop diameters depend on the intensity of agitation and the physical/transport properties of the continuous and disperse phases. However, these calculations are limited to very low disperse phase viscosities and holdup fractions. Doulah [66] proposed a correction to the derived correlations to account for high holdup disperse volume fractions, whereas Arai et al. [77] derived an expression for the maximum droplet diameter by incorporating the viscosity of the disperse phase. Similar expressions were also proposed by Calabrese et al. [69].

Several models have been proposed for calculation of the drop breakage rate in liquid–liquid dispersions [63, 68, 78–81]. Some of these models have been applied to the suspension polymerization process with great success [25, 58, 78, 81]. In these models, the drop breakage rate is expressed in terms of breakage frequency, $\omega_b(v)$, and a respective Maxwellian efficiency term:

$$g(v) = \omega_b(v)e^{-\lambda_b(v)} \tag{66}$$

where $\lambda_b(v)$ is the ratio of the energy required for drop breakage to occur to the available energy.

Let us assume that a drop of volume $u$ breaks up into $N_{da}$ daughter drops and $N_{sa}$ satellite drops. Furthermore, let us assume that the daughter and satellite drops are normally distributed about their respective mean values, $v_{da}$ and $v_{sa}$. One can derive the following expression for the number of drops of volume $v$ formed by the breakage of a drop of volume $u$ [82]:

$$\beta(u,v)\, u(u) = N_{da}\left\{\frac{1}{\sigma_{da}\sqrt{2\pi}}\exp\left(-\frac{(v-v_{da})^2}{2\sigma_{da}^2}\right)\right\}$$
$$+ N_{sa}\left\{\frac{1}{\sigma_{sa}\sqrt{2\pi}}\exp\left(-\frac{(v-v_{sa})^2}{2\sigma_{sa}^2}\right)\right\} \tag{67}$$

It should be noted that the daughter drop number density function, $u(u)\beta(u,v)$, should satisfy the following number and volume conservation equations:

$$\int_0^u u(u)\,\beta(u,v)\,dv = u(u), \quad \int_0^u v\,u(u)\beta(u,v)dv = u \tag{68}$$

Accordingly, one can calculate the mean volumes of daughter and satellite drops formed by the breakage of a drop of volume $u$ in terms of $N_{da}$ and $N_{sa}$ and the ratio of their respective volumes, $r_D = v_{da}/v_{sa}$ [48]:

$$v_{da} = \frac{u}{N_{da}+N_{sa}/r_D} \quad \text{and} \quad v_{sa} = \frac{u}{N_{da}\,r_D+N_{sa}} \tag{69}$$

where $N_{da}$, $N_{sa}$, $\sigma_{da}$, $\sigma_{sa}$, and $r_D$ are model parameters.


## 3.4 Droplet Coalescence Process

Two different mechanisms have been postulated in the literature to describe the coalescence of two drops in a turbulent flow field. The first assumes that, after the initial collision of two drops, a liquid film of the continuous phase is trapped between the two drops, preventing their coalescence [61]. However, the presence of attractive forces leads to draining of the liquid film and drop coalescence. On the other hand, if the kinetic energy of the induced drop oscillations is larger than the energy of adhesion between the drops, then drop contact is broken before complete drainage of the liquid film. The second drop coalescence mechanism [86] assumes that immediate coalescence occurs when the relative velocity of the two colliding drops at the collision instant exceeds a critical value. This means that the drops coalesce if the energy of collision is greater than the total drop surface energy.

Several mathematical models have been presented to describe the drop coalescence rate [67, 78, 79, 83–90]. As in the case of the drop breakage rate, the drop coalescence rate can be expressed in terms of collision frequency, $\omega_b(v,u)$, and a Maxwellian efficiency term:

$$k(v, u) = \omega_c(v, u)\, e^{-\lambda_c(v,u)} \tag{70}$$

where $\lambda_c(v,u)$ is the ratio of the energy required for drop coalescence to the available energy.

Detailed expressions for calculation of the drop breakage and coalescence rate kernels can be found in the original publication of Kotoulas and Kiparissides [25].

## 3.5 Physical and Transport Properties

One of the most important issues in modeling the suspension polymerization process is evaluation of the physical and transport properties of the reacting system, as well as calculation of partitioning of the different species (monomer(s), polymer, initiator(s), etc.) in the various phases present in the system. In a suspension polymerization process, one can identify at least three phases: the disperse phase (e.g., polymerizing monomer droplets), the continuous aqueous phase, and the gas phase. The disperse phase can be either homogeneous (if the polymer is soluble in its monomer) or heterogeneous (if the polymer is insoluble in its monomer). In powder suspension polymerization, the disperse phase consists of two different phases, polymer-rich and monomer-rich. The continuous aqueous phase contains only small amounts of monomer and the gas phase contains monomer and water vapors. The density of the suspension system, $\rho_s$, can be calculated from the weighted sum of the densities of the disperse ($\rho_d$) and continuous ($\rho_c$) phases [91]:

$$\rho_s = \rho_d \varphi + \rho_c (1 - \varphi) \tag{71}$$

where $\varphi$ is the volume fraction of the disperse phase. The density of the polymerizing monomer droplets (disperse phase) is in turn a function of the polymer ($\rho_p$) and monomer ($\rho_m$) densities and the extent of monomer conversion, $X$:

$$\rho_d = \left( \frac{X}{\rho_p} + \frac{1-X}{\rho_m} \right)^{-1} \tag{72}$$

Accordingly, the viscosity of the dispersion system can be calculated by the following semi-empirical equation [92]:

$$\eta_s = \frac{\eta_c}{1 - \varphi} \left( 1 + \frac{1.5 \eta_d \, \varphi}{\eta_d + \eta_c} \right) \tag{73}$$

where $\eta_d$ and $\eta_c$ are the viscosities of the disperse and continuous phases, respectively.

For the suspension polymerization of VCM, the viscosity of the polymerizing monomer droplets, $\mu_d$, can be calculated from the Eilers equation [93]:

$$\eta_d = \eta_m \left( 1 + \frac{0.5[\eta]_{\text{pol}}}{1 - \varphi_{\text{pol}}/\varphi_{\text{cr}}} \right)^2 \tag{74}$$

where $\varphi_{\text{pol}}$ is the volume fraction of the polymer in the dispersed phase, given by $\varphi_{\text{pol}} = X(\rho_d/\rho_{\text{pol}})$. Term $\varphi_{\text{cr}}$ is the polymer volume fraction corresponding to the critical monomer conversion, $X_c$, at which a 3D polymer skeleton is formed inside the polymerizing monomer drops. When $\varphi_{\text{pol}}$ approaches the $\varphi_{\text{cr}}$ value, the dispersed-phase viscosity approaches a limiting constant value, corresponding to a rigid structure. The value of $\varphi_{\text{cr}}$ for VCM suspension polymerization was taken as 0.3 [59].

The intrinsic viscosity of the polymer solution, $[\eta]_{\text{pol}}$ in Eq. (74) can be calculated by the well-known Mark–Houwink–Sakurada (MHS) equation as a function of polymer $M_w$:

$$[\eta]_{\text{pol}} = k \, M_w^a \tag{75}$$

Finally, the viscosity of the continuous phase depends on the concentration and type of stabilizer, which, in turn, affects the particle size distribution [54]. Okaya [94] employed the following Schulz–Blaschke equation to calculate the viscosity of aqueous PVA solutions:

$$\eta_c = \eta_w \left( 1 + \frac{[\eta_{\text{PVA}}]C_{\text{PVA}}}{1 - 0.45[\eta_{\text{PVA}}]C_{\text{PVA}}} \right) \tag{76}$$

where $\mu_c$, $\mu_w$, $[\eta_{\text{PVA}}]$, and $C_{\text{PVA}}$ are the viscosities of the aqueous PVA solution and pure water, and the intrinsic viscosity and concentration of the stabilizer, respectively. A great number of papers have been published dealing with the behavior of polymer molecules at interfaces. Prigogine and coworkers [95] presented a remarkably simple theory on the calculation of surface tension of polymer solutions.

Although the Prigogine theory refers specifically to the surface tension of polymer solutions, it is equally applicable to prediction of interfacial tension between a polymer solution and an immiscible liquid or a solid [96]. In the present study, the model of Siow and Patterson [96] was employed for calculation of the interfacial tension between aqueous and dispersed phases. The change in interfacial tension with monomer conversion was taken into account, as in the original work of Maggioris et al. [58].

## 3.6 Numerical Solution of the Population Balance Equation

In general, the numerical solution of the dynamic PBEs for a particulate process, especially for a polymerization, is a difficult problem because of both numerical complexities and model uncertainties regarding particle growth, aggregation, and breakage mechanisms that are often poorly understood. Usually, the numerical solution of a PBE requires discretization of the particle volume domain into a number of discrete elements, which results in a system of stiff, nonlinear differential or algebraic/differential equations that are solved numerically [97]. Several numerical methods have been developed for solving steady-state or dynamic PBEs. These include the full discrete method [98], the method of classes [48, 99], the discretized PBE [100, 101], fixed and moving pivot techniques [60, 102], high-order discretized PBE methods [103–105], orthogonal collocation on finite elements [106], the Galerkin method [107], and the wavelet-Galerkin method [108]. The reviews of Ramkrishna [109] and Dafniotis [110] describe the various numerical methods available for solving PBEs in detail. Moreover, the publications of Kiparissides and coworkers [111–114] present comparative studies on the different numerical methods.

The numerical solution of PBEs commonly requires discretization of the particle volume domain into a number of discrete elements. Accordingly, the unknown number density function is approximated at a selected number of discrete points, resulting in a system of stiff, nonlinear differential equations that are subsequently integrated numerically. In the present work, the fixed pivot technique [25, 97, 102] was employed for solution of the resulting PBE [67]. Assuming that the number density function remains constant in the discrete volume interval ($v_i$ to $v_{i+1}$), one can define a particle number distribution, $N_i(t)$, corresponding to the $i$ element:

$$N_i(t) = \int_{v_i}^{v_{i+1}} n(v, t)\, dv = \bar{n}_i(v, t)(v_{i+1} - v_i) \qquad (77)$$

Following the original developments of Kumar and Ramkrishna [102], the total volume domain ($v_{min}$ to $v_{max}$) is first divided into a number of elements. The drop/particle population, $N_i(t)$, corresponding to the size range ($v_i$, $v_{i+1}$) is then assigned to a characteristic size $x_i$ (also called the grid point). The accuracy and convergence characteristics of the numerical method were first assessed by varying the total number of discretization points, the size of the total volume domain, and the initial droplet size distribution. The diameter domain extended from 1 to 2,000 μm and the initial droplet size distribution, $D_o$, followed a Gaussian distribution with a mean value of 1,000 μm (standard deviation $\sigma_D = 100$ μm). The volume probability density function converged to the same distribution for number of elements ≥80. In the present study, it was assumed that the numerically calculated distribution converged to the correct value when the total mass of the dispersed phase (i.e., monomer plus polymer), given by the first moment of particle number distribution, differed from the initial monomer mass by less than 2%.

## 3.7  Effect of Operating Conditions on Particle Size Distribution

An increase in the input power per unit mass (e.g., increasing the agitation speed or the impeller diameter) causes an increase in the turbulent intensity and fluctuations in pressure and velocity. As a result, the drop breakage rate increases, leading to the production of smaller and more uniform polymerizing droplets. At the same time, the increased liquid circulation rate results in more drop collisions, which increases the drop coalescence rate. In general, an increase in input power to the system results in an increase in drop breakage and coalescence rates. Depending on which of two (i.e., drop breakage or drop coalescence) mechanisms dominates, an increase in input power could lead to a shift of the mean drop diameter to lower or higher values. Moreover, it has been observed that the mean drop/particle diameter follows a U-shape variation with respect to impeller speed [17, 52] and impeller diameter [115].

A large number of experimental and theoretical studies have been published on the effect of continuous and disperse phase viscosity on the PSD. In general, an increase in disperse phase viscosity, $\eta_d$, results in a reduction of both breakage and coalescence rates. Cebollada et al. [54] reported that, in the suspension polymerization of VCM, increasing the viscosity of the continuous phase resulted in production of larger and more uniform polymer particles. On the other hand, as the viscosity of the continuous phase decreased, the PVC sub-grains were smaller in size but their agglomeration rate increased. As a result, larger grains with higher porosity were produced.

In general, an increase in holdup fraction of the disperse phase, $\varphi$, decreases the turbulent intensity (i.e., the average energy dissipation rate per unit mass) and, thus, the drop breakage rate. The coalescence frequency increases as a result of the higher number of droplets, while the coalescence efficiency decreases because of the lower average energy dissipation rate. However, the effect of $\varphi$ on the

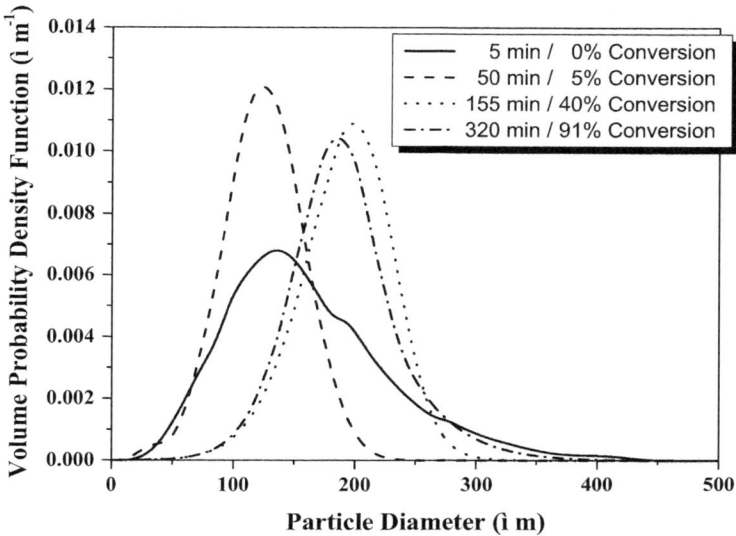

**Fig. 15** Dynamic evolution of PSD with respect to polymerization time for VCM suspension polymerization; polymerization temperature 56.5°C, impeller speed 330 rpm, disperse phase volume fraction 40%

coalescence frequency is more important; thus, for a constant input power, the droplet size increases as the holdup fraction increases [74].

Etesami et al. [116] investigated experimentally the effect of the phase ratio (VCM/water) on the particle properties of PVC resins produced by suspension polymerization. They reported that an increase in $\varphi$ resulted in broader and multimodal PSD. The mean particle size and the bulk density of the PVC resin also increased with $\varphi$, while the grain porosity decreased. On the other hand, the average molecular weights and polydispersity index did not change with $\varphi$.

Figures 15 and 16 illustrate the dynamic evolution of the PSD and Sauter mean particle diameter, respectively. More specifically, in Fig. 15 the volume probability density function [defined as $v\, n(v,t)/V_{tot}$] of PVC particles is plotted with respect to particle diameter. In these figures, one can easily distinguish the three stages of the suspension polymerization process. For VCM suspension polymerization, the PSD is essentially established at monomer conversions of about 35–40%.

## 3.8 Effect of Overhead Condenser Operation Mode on Particle Size Distribution

One of the most efficient ways to remove polymerization heat is to use an overhead reflux condenser. The VCM vapors are condensed and return to the polymerizing suspension, while an equal amount of VCM is vaporized to maintain the

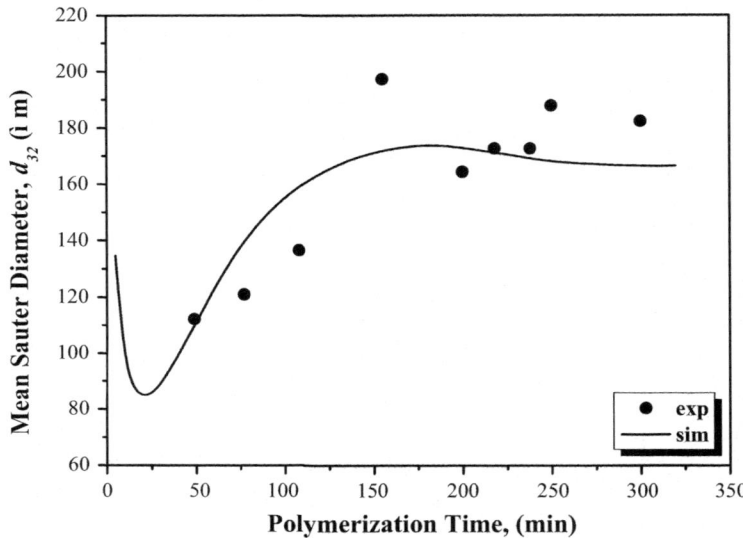

**Fig. 16** Dynamic evolution of the Sauter mean diameter of PVC particles with respect to polymerization time (experimental conditions same as for Fig. 15)

thermodynamic equilibrium. However, this process affects the morphological properties of the polymer product (e.g., porosity, PSD, bulk density). Cheng and Langsam [53] reported that the particle size and porosity of PVC grains increase as the operation time of the condenser or/and the reflux rate increase. This result has also been verified by Zerfa and Brooks [74], who observed a second peak in the PSD at higher sizes as the reflux rate increased while, at the same time, some fine particles were produced. The large peak corresponds to the PVC particles produced either by polymerization of the larger "fresh" monomer droplets returning from the condenser or by the coalescence of fresh and "old" droplets. Fresh droplets are larger than old droplets because the stabilizer concentration in the continuous phase cannot sufficiently cover the newly formed monomer droplets. The operation time of the condenser is another important factor in the suspension polymerization of VCM. Cheng and Langsam [53] reported that the reflux condenser should only be used after a monomer conversion of about 5%, because early utilization of the reflux condenser results in extensive condenser fouling and a coarser resin. On the other hand, if utilization of the condenser starts at high monomer conversions (20–30%), its effect on the PSD is negligible although its effect on grain porosity remains significant.

Etesami et al. [117, 118] investigated experimentally the effect of reflux rate during suspension polymerization on the particle properties of PVC resin. It was found that the monomer conversion decreased with increasing reflux rate. It was also observed that the cold plasticizer absorption increased with reflux rate, whereas the bulk density and $K$ value of the resin decreased. Scanning electron microscopy

(SEM) micrographs showed that PVC resin with a rougher particle surface, more separate aggregates, and smaller primary particles was prepared at higher reflux rates. It was also found that commencement of refluxing before 20% conversion resulted in bimodal PSD, whereas monomodal PSD was obtained with longer delays in refluxing.

It should be noted that utilization of a reflux condenser for heat removal in suspension polymerization of VCM introduces some operational problems. For example, noncondensable gases may become concentrated in the condenser and reduce its heat removal capacity. The amount of noncondensable gases in the reactor overhead vapor phase depends on the quality of the monomer, how well the reactor has been evacuated prior to polymerization, and whether or not the polymerization process generates inert gases [53]. For example, the use of azo initiators results in the formation of nitrogen from the decomposition of initiator molecules. In addition, the use of a carbonate buffer can result in the formation of $CO_2$ if the aqueous phase becomes acidic.

## 3.9   Scale-up of Suspension Polymerization Reactors

The scale-up of suspension polymerization reactors (i.e., from laboratory to pilot and then to industrial scale) is not straightforward or well established. Probably, the most significant problem in scale-up occurs when different physical processes become limiting at different scales. For example, commercial-scale suspension reactors have to perform several functions simultaneously (dispersion, reaction, and heat transfer), which do not scale up in the same manner. Thus, heat removal can become a limiting factor for reactor performance at large scales, whereas it is rarely a problem for laboratory-scale reactors [119].

In suspension polymerization, scale-up of an agitated tank reactor should not change the particle morphology (e.g., particle size distribution, porosity, bulk density) of the polymer product, given that the polymerization recipe and operating conditions are kept constant, and the reactor design can accommodate removal of the generated polymerization heat. Thus, the problem is reduced to the scale-up of a liquid–liquid dispersion in agitated vessels. This can be addressed on the basis of several particle size distribution criteria, including constant power input per unit volume, impeller discharge flow rate, impeller tip speed, Weber number, and Reynolds number. [120]. Assuming constant reactor geometry at scale-up, the criterion of constant power input per unit mass yields the following equation:

$$N^3 D_I^2 = \text{constant}; \quad N D_I^{0.66} = \text{constant} \tag{78}$$

Equation (78) assumes that the power number, $N_p$, remains constant. The power number represents the ratio of pressure to inertia forces [121]. The above criterion assumes dynamic similarity conditions for large Reynolds numbers (i.e., negligible viscous forces) and no effect of gravitational forces. The criterion of equal impeller tip speed, under constant reactor geometry, leads to the relation $N D_I = $ constant.

Okufi et al. [121] studied the effect of vessel size (i.e., scale-up) on the droplet size distribution of $n$-heptane in water. They reported that the rule of equal impeller tip speed provided the best scale-up criterion, assuming a constant interfacial area per unit volume for liquid–liquid dispersions.

Scully [122] used the well-known correlation between the Sauter mean particle diameter, $d_{32}$, and the Weber number $\left(N_{We} = N^2 D_I^3 \rho / \sigma\right)$ to derive a scale-up criterion for suspension PVC reactors:

$$d_{32}/D_I = C N_{We}^{-0.6} \tag{79}$$

Equation (79) does not account for the viscous forces inside the polymerizing droplet.

Calabrese et al. [69] proposed the following relation for the calculation of the Sauter mean droplet diameter for a viscous dispersion system:

$$d_{32}/D_I = 0.0053 \left(1 + 0.91 \text{Vis}^{0.84}\right)^{0.6} N_{We}^{-0.6} \tag{80}$$

where

$$\text{Vis} = \left(\rho_c/\rho_d\right)^{1/2} \left(\mu_d N D_I / \sigma\right) \tag{81}$$

Equation (80) can be used as a scale-up criterion to produce polymer particles with the same Sauter mean diameter. In this case, the criterion derived is:

$$N D_I^{0.43} = \text{constant} \tag{82}$$

Lewis and Johnson [123] studied experimentally the effects of agitation intensity and reactor size on the mean particle size and the bulk density of PVC grains. Experiments were carried out in three stirred batch polymerization reactors of different vessel sizes (bench scale, pilot plant, and commercial production units). The three reactors were geometrically similar. The same polymerization recipe and operating conditions were used for all three reactors. The effects of major agitation parameters such as impeller diameter, width, and speed were correlated against resin properties using the Weber number. The same characteristic U-shaped curve was found for all three reactors when the mean particle diameter was plotted against Weber number. The three U-shaped curves did not lie on top of each another but were spread apart. As the reactor size increased, the value of the Weber number corresponding to the minimum in the U-shape curve (i.e., minimum mean particle size) shifted to larger values. However, the results indicated that the minimum particle size was not affected by scale-up. Another interesting observation was that the coefficient of variation (the particle size standard deviation divided by the mean particle diameter) decreased dramatically as the reactor size increased. Three different correlations of the bulk density of the resin with respect to the Weber number were established for the three reactor sizes. In all cases, it was found that the bulk density was quite high (e.g., 0.7 g/cm$^3$) at low values of the Weber number. However, as the Weber number increased, the bulk density decreased, leveling off to an approximate value of 0.5 g/cm$^3$. At even higher Weber numbers, the bulk density again decreased. Note that, as the reactor size increased, the bulk density leveled off at slightly higher levels.

Ozkaya et al. [124] studied experimentally the suspension polymerization of VCM at different reactor scales (i.e., from 10 L to 27 m$^3$). They found that the mean particle size, $d_{50}$, depended on the Weber number according to the following relation:

$$d_{50}/D_I = 2.73 \times 10^5 (N_{We})^{-0.51} \tag{83}$$

The scale-up criterion that can be derived from Eq. (83), for geometric similarity of the reactors, is:

$$N D_I^{0.52} = \text{constant} \tag{84}$$

# 4 Calculation of Grain Morphology

In general, PVC grains consist of a number of sub-grains (agglomerated droplets), depending on the quality of agitation and the stability of the VCM droplets [125]. Thus, under low intensity agitation conditions and moderate values of monomer–water interfacial tension, unicellular grains can be produced consisting of finer (i.e., about 50 μm) and denser (i.e., low porosity) polymer particles. On the other hand, multicellular grains, having an average size of 100–150 μm, can be produced under more intense agitation conditions and lower values of interfacial tension. The grain porosity can vary significantly between uni- and multicellular PVC grains. For multicellular grains, in addition to the intracellular porosity of the unicellular grains, macropores formed between the agglomerated droplets can give rise to the so-called intercellular porosity.

It has been reported that PVC grains are frequently surrounded by an outer pericellular membrane of approximately 200–500 nm in thickness [126]. Initially, a thin polymer "skin" is formed at the VCM–water interface as a result of grafting of PVC chains onto the absorbed primary stabilizer molecules. Subsequently, primary particles formed within the disperse monomer droplets aggregate at the polymer skin, leading to the formation of a pericellular membrane [127]. The morphology (i.e., porosity) of the pericellular membrane depends on the type and concentration of the primary stabilizer and affects the droplet coalescence and breakage processes, as well as the extent of droplet/particle volume contraction caused by the higher density of the polymer [128]. These effects, however, have not been sufficiently studied nor quantified in the literature. Note that the porosity of the pericellular membrane is directly related to the accessibility of internal grain pores and, thus, to plasticizer uptake by PVC grains. Moreover, when the pericellular membrane pores are closed, a pressure difference across the membrane develops during the course of polymerization. This can cause the internal particle network to "collapse," resulting in the formation of "dimpled" particles with concomitant the loss of grain porosity [129].

In general, the morphological properties of PVC grains are determined by the following process variables: polymerization temperature, quality of agitation, and type and concentration of surface-active agents (i.e., primary and secondary stabilizers). The relationships between process variables (agitation, temperature, stabilizers, etc.), microscale grain characteristics (PPSD, porosity, etc.), and macroscale grain size distribution are not well-understood because many complex interrelated physical and chemical phenomena are involved [6]. Because of this multiscale complexity, it is not surprising that there are only a few comprehensive models accounting for the effect of process variables on grain porosity [26, 130]. Previously published papers [6, 8, 23, 26, 129–139] on the morphology of PVC grains have postulated a five-stage kinetic-physical mechanism (as shown in Fig. 11) to describe the nucleation, stabilization, growth, and aggregation of PVC primary particles.

In the following section, a comprehensive population balance model is developed to calculate the dynamic evolution of the PPSD in polymerizing monomer droplets in terms of the process variables. Model predictions on the total number of primary particles, average primary particle diameter, PPSD, and the critical monomer conversion at which the primary particles form a 3D polymer network inside the polymerizing monomer droplets, are compared with available experimental measurements.

## 4.1   PVC Primary Particle Size Distribution

The PPSD is a very important property because it largely controls the porosity of the final PVC grains. The dynamic evolution of the PPSD is controlled by many process variables, including polymerization temperature, ionic strength of the medium, and type and concentration of secondary stabilizer. The total number and average size of PVC primary particles have been experimentally measured at different monomer conversions (see Table 5). The results are somehow conflicting because of variations in the polymerization conditions, especially the agitation rate, stabilizer type, and stabilizer concentration.

Smallwood [126] conducted an experimental investigation of the structure of suspension PVC grains up to high monomer conversions (5–85%), at different temperatures (51–71°C) and agitation rates (175–225 rpm). Under those conditions, the measured primary particle size and number were approximately 1.4 μm and

**Table 5** Summary of experimental data on PVC primary particles

| Authors | Type | Initiator[a] | Temperature (°C) | Agitation rate (rpm) | Stabilizers | Range | Measured |
|---|---|---|---|---|---|---|---|
| Willmouth et al. [134] and Rance and Zichy [138] | Bulk, 0.15 mL VC $D_T = 2.5$–3 mm | LP 1 g/kg | 35–50 | 0 | None | $t = 15$–50 min; $x < 0.4\%$ | Particle size and number |
| Tornell et al. [137] | Bulk $V_T = 250$ mL | LP | 60 | 0 | None | $x = 0.5\%$ | PSD (at $x = 0.5\%$) |
| Tornell and Uustalu [136][b] | Bulk, 120 g VC $D_T = 52$ mm, $D_I = 48$ mm | CEPC | 40 | 0–0.4–1.5 m/s | None | $x = 0.1$–3% | Particle size and number; porosity and bulk density |
| Tornell and Uustalu [135] | Bulk, 120 g VCM; $D_T = 52$ mm, $D_I = 48$ mm | 0.11 g LP | 59 | 159–597 | Span20, Span80, Tween21, PMMA | $x = 0$–8% | Particle size and number; PSD (at $x = 2$ and 7%) |
| Boissel and Fischer [141] | Bulk $V_T = 1$ L, $D_I = 62$ mm | LP 3.2 g/kg | 30–80 | 250–1,200 | None | $t = 120$ min; $x < 0.6\%$ | Particle size and number |
| Davidson and Witenhafer [127] | Suspension 10 mL VCM, 20 mL water; $D_T = 20$ mm | ACHS 0.3% (w/v) | – | Yes | PVA 0.25 g/kg | $x = 2$–4% | SEMs and micrographs of PVC grains |
| Nilsson et al. [56] | Suspension 6.5 L water, 5.1 kg VCM | CEPC 3.1 g | 55 | Yes | Gohsenol GH20, Methocel F50, Rhodoviol 5/270, Span20, AL | $x = 0$ to >75% | DSD $x = 5\%$; PSD $x = 100\%$; porosity |
| Smallwood [126] | Suspension $V_T = 160$ L | LP and APC | 51–71 | 175–225 | PVA 72.5, cellulose | $x = 5$–80% | Porosity, grain size; particle size and number; SEM micrographs of grains |

[a]*LP* lauryl peroxide, *CEPC* dicetyl peroxydicarbonate, *ACHS* acetyl cyclohexyl sulfonyl peroxide, *APC* alkyl peroxydicarbonate, *AL* ammonium laurate, *t* polymerization time, *x* percentage monomer conversion, $D_T$ reactor diameter, $D_I$ impeller diameter, $V_T$ the reactor volume

[b]Addition of $Bu_4NBF_4$ and $Bu_4NCl$ electrolytes at 0.02–2 wt% concentration

$2.0 \times 10^{11}$ cm$^{-3}$, respectively. Smallwood found that an increase in polymerization temperature caused a significant decrease in grain porosity, apparently due to particle fusion (evident in SEM micrographs). Moreover, he reported that the agitation rate did not significantly affect the primary particle size. However, the range of agitation rates examined was rather limited.

Willmouth et al. [134] studied bulk VCM polymerization at temperatures ranging from 35 to 60 °C, and low monomer conversions (e.g., 0.25%). They reported that, at very low monomer conversions, the primary particle nucleation rate was larger than the particle aggregation rate. Thus, initially both the total particle number and the particle size increased with time. However, at higher monomer conversions, the particle coagulation rate became the dominant mechanism and the total particle number started decreasing linearly with time. During this stage, the primary particle diameter increased.

Tornell and Uustalu [131, 135, 136] studied the effects of agitation rate and addition of various secondary stabilizers and additives [Span20, Tween21 and poly (methyl methacrylate)] on the PPSD for the suspension polymerization of VCM. They found that, in the presence of Span20 stabilizer, the total number of primary particles increased and the average particle diameter decreased as the stabilizer concentration increased. They also reported that, in the presence of Span20, massive particle aggregation (i.e., formation of a 3D polymer network) occurred at lower monomer conversions (i.e., <7%).

Tetrabutyl ammonium tetrafluoroborate and tetrabutyl ammonium chloride dissociate completely in VCM. These salts, when added to colloidal dispersions of PVC primary particles in VCM prepared by bulk polymerization, caused the particles to flocculate, showing that the primary particles are electrostatically stabilized. When bulk polymerizations were carried out in the presence of the same quarternary ammonium salts, smaller primary particles were obtained than in additive-free polymerizations. Addition of tetrabutylammonium tetrafluoroborate to a suspension polymerization increased the porosity of the resin [136].

The stability of primary particles is generally assumed to be of electrostatic nature because of the formation of chlorine ions during VCM polymerization [137]. However, in the presence of secondary stabilizers, the steric stabilization of primary particles can also be important. One possible source of chlorine ions is the production of HCl in free-radical VCM polymerization. Rance and Zichy [138] estimated that the number of negative charges for primary particles of 80 nm in diameter was 41, that is, 2–3 orders of magnitude lower than the reported values for aqueous latexes. From zeta-potential measurements, the primary particle potential was found to vary from −80 to −120 mV [127, 134, 139]. Willmouth et al. [134] estimated that the total interaction potential between primary particles of 100 nm was about 10–13 $kT$, which is smaller than the interaction potentials measured in well-stabilized latex dispersions. Despite the small interaction potential, Davidson and Witenhafer [127] observed that, in quiescent polymerization conditions, primary particles appear to form regular arrays indicating a long-range stabilization force.

Tornell et al. [137] derived a two-parameter electrostatic stabilization model for primary PVC particles. A Debye double-layer thickness of $\kappa^{-1} = 100$ nm was estimated, corresponding to an ionic concentration of 1 μmol/L of $BF_4^-$. DLVO theory [140] was applied to determine interaction potentials at constant particle surface potential and constant particle surface charge. Particle destabilization and massive aggregation of larger particles was postulated to occur either by a decrease in the maximum interaction potential caused by dilution of surface charges, or by shear induced aggregation. The limitations of classical DLVO theory were also noted. For example, the surface charge density and counterion concentration may be too small to correspond to a continuous distribution of charges as required by the DLVO theory.

### 4.1.1 Calculation of Primary Particle Size Distribution

Despite the importance of the PPSD on the development of PVC grain morphology, there are only a limited number of quantitative models dealing with the dynamic evolution of primary particles in terms of process variables [13, 26, 59, 142]. Alexopoulos and Kiparissides [26] developed a PBE model to describe dynamic evolution of the PPSD in terms of polymerization temperature and ionic strength of the reaction medium by extending the electrostatic DLVO theory to the charged primary particles of Tornell et al. [137]. The PBE included nucleation, aggregation, and growth terms and was solved using the method of orthogonal collocation on finite elements.

In general, dynamic evolution of the PSD in a particulate process can be obtained from the solution of a PBE [109, 143]. Thus, the population of the primary particles can be expressed in terms of a number density function, $n(v, t)$, that represents the number of particles per unit volume of monomer droplet phase in the differential size range ($v$ to $v+dv$). For a dynamic system undergoing particle nucleation, aggregation, and growth, the evolution of the PSD is given by the following nonlinear integro-differential PBE:

$$\frac{dn(v,t)}{dt} + \frac{d(G(v,t)n(v,t))}{dv} + n(v,t)\frac{d(\ln(V_d(t)))}{dt} = \delta(v-v_0)S_0(t)$$

$$+ \int_0^{v/2} \beta(v-u,u)n(v\text{-}u,t)n(u,t)du - \int_0^{v_{max}} \beta(v,u)n(v,t)n(u,t)du \tag{85}$$

where $G(v,t)$ is the particle growth rate due to polymerization in the polymer-rich phase, $S_0(t)$ is the nucleation rate of primary particles of volume $v_0$ in the monomer-rich phase, and $\beta(u,v)$ is the aggregation rate kernel for particles of volume $u$ and $v$. It should be noted that due to the strong attractive forces between particles, breakage is not considered in Eq. (85). In general, Eq. (85) satisfies the following initial and boundary conditions:

$$n(v,0) = 0 \quad \text{at} \quad t = 0; \qquad n(0,t) = 0 \quad \text{at} \quad v = 0 \tag{86}$$

The third term on the right hand side of Eq. (85) represents the effect of monomer drop shrinkage during polymerization. To solve Eq. (85), the functional forms of the particle growth rate, $G(v,t)$, particle nucleation rate, $S_0(t)$, and the particle aggregation rate kernel, $\beta(u,v)$, need to be first determined.

### 4.1.2 Nucleation and Growth Rates of Primary Particles

As discussed in detail in the previous section, polymerization in the monomer-rich phase results in the formation of PVC domains (i.e., primary particle nuclei). According to Kiparissides [144], the primary particle nucleation and growth rates is given by:

$$S_0(t) = \frac{R_{pm}M_w}{\rho_p(1 - \varphi_m)v_o^2} \tag{87}$$

$$G(v,t) = \frac{M_w R_{pp}}{\rho_m X} v \tag{88}$$

where $X$ is the monomer conversion, $\varphi_m$ is the volume fraction of monomer in the polymer-rich phase, $M_w$ is the molecular weight of VCM, $\rho_m$ and $\rho_p$ are the corresponding monomer and polymer densities. $R_{pm}$ and $R_{pp}$ denote the respective polymerization rates in the monomer-rich and polymer-rich phases. Note that the growth rate is linear with respect to the particle volume, which is typical of bulk polymerization systems, and depends on $R_{pp}$ and, thus, on time.

$R_{pm}$ and $R_{pp}$ can be calculated from a kinetic model [6, 26, 144–146]. In the present study, for simplicity, the kinetic model of Abdel-Alim and Hamielec [11] was employed for the calculation of the $S_0(t)$ and $G(v,t)$ functions:

$$R_{pp} = k_I[M][I]^{1/2}P\frac{(1 - X_f)X}{X_f(1 - X)} \tag{89}$$

$$R_{pm} = k_I[M][I]^{1/2}\frac{(1 - BX_f)}{X_f}\frac{(X_f - X)}{X_f(1 - X)} \tag{90}$$

where $k_I$ is the rate constant for initiator decomposition, $[M]$ is the monomer concentration, $[I]$ is the initiator concentration, and $X_f$ is the VCM conversion at which the separate monomer-phase disappears. Finally, the rate of change of monomer conversion is given by:

$$\frac{dX}{dt} = k_p\sqrt{2fk_d/k_t}(1 - X - AX + PAX)[I_0]^{1/2}\exp(-k_dt/2)/\sqrt{1 - BX} \tag{91}$$

**Table 6** Dimensionless coefficients of the kinetic model

$$B = (\rho_p - \rho_m)/\rho_m$$

$$A = (1 - X_f)/X_f$$

$$P = \left(\sqrt{2fk_d/k_t}\right)_p / \left(\sqrt{2fk_d/k_t}\right)_m \approx 27 - 0.14\,T\ (^\circ C)$$

where $k_p$, $k_d$, and $k_t$ are the rate coefficients for monomer phase propagation, initiator dissociation, and termination, respectively. The temperature-dependent dimensionless parameters $A$, $B$, and $P$ are given in Table 6.

### 4.1.3 Aggregation Rate of Primary Particles

In VCM polymerization, the total number of particles and evolution of the PPSD are controlled by the nucleation and aggregation rates of the primary particles. During the initial stages of polymerization, particle aggregation follows mainly a perikinetic mechanism (i.e., diffusion-driven). Thus, the aggregation rate between two colloidal particles of radii $r_i$ and $r_j$ can be expressed by the modified Smoluchowski equation [147]:

$$\beta_{ij} \equiv \beta(r_i, r_j) = \frac{2k_B T}{3\mu} \frac{(r_i + r_j)^2}{r_i r_j} \frac{1}{W_{ij}} \tag{92}$$

where $k_B$, $T$, $\mu$, and $W_{ij}$ denote the Boltzmann constant, reaction temperature, viscosity of the continuous monomer phase, and the Fuch's stability ratio, respectively. The stability ratio, $W_{ij}$, relates the actual aggregation rate to the uncontrolled fast Smoluchowski aggregation rate. Calculation of the stability ratio is described in the original publication of Alexopoulos and Kiparissides [26].

To take into account the effects of agitation rate, the concentrations of primary and secondary stabilizers and initiator on the stability of the primary particles and, thus, on the value of $X_c$, the stability ratio needs to be properly modified. Due to the complexity of the problem and the lack of a sufficient amount of experimental data, first-principle models cannot be pursued. Instead, phenomenological and/or semi-empirical models were developed. The parameters associated with these models were determined by using available experimental data on the critical monomer conversion, $X_c$, and the PPSD.

### Effect of Agitation

The effect of agitation on critical monomer conversion and, consequently, on grain porosity has been clearly demonstrated in the bulk polymerization experiments of Boissel and Fischer [141] and Davidson and Witenhafer [127]. However, this effect is expected to be less pronounced in suspension polymerization because of the "shielding" effect of the pericellular membrane of VCM droplets. Even so, an increase in grain porosity with increased agitation rate has been reported in the literature for the suspension polymerization of VCM [126]. The mechanism by which an increase in the agitation rate causes the primary particles to aggregate earlier is thought to be due to shear-induced aggregation.

During the early stages of polymerization, the aggregation of small primary particles is governed by a diffusion-driven mechanism (perikinetic). However, as the size of the primary particles increases, shear-induced aggregation becomes increasingly important (orthokinetic). The primary particle fluxes for perikinetic and orthokinetic aggregation is given by [150]:

$$J_P = \frac{2}{3} \frac{k_B T}{\mu} \frac{(r_1 + r_2)^2}{r_1 r_2} \tag{93}$$

$$J_O = \frac{4}{3} \dot{\gamma} (r_1 + r_2)^3 \tag{94}$$

The ratio $J_O/J_P$ is effectively a Peclet number and depends on the mean value of the shear rate, $\dot{\gamma}$. Based on the third-order dependence of $J_O$ on the particle radii in Eq. (94), it is clear that the orthokinetic mechanism dominates the aggregation of larger particles.

The two aggregative particle fluxes $J_P$ and $J_O$ are not directly additive, and analytical solutions for combined orthokinetic and perikinetic aggregation mechanisms are not available. However, an approximate solution exists for the case where perikinetic aggregation is the dominant mechanism [148, 149]. By modifying the proposed approximate solution, the combined orthokinetic-perikinetic aggregation rate kernel $\beta^{OP}_{ij}$ between particles of radii $r_i$ and $r_j$ can be obtained:

$$\beta^{OP}_{ij} = \frac{1}{W_{ij}} \left( J_P + 0.4(J_P J_O)^{1/2}/W_{ij} + C_1 (J_O)^{C_2} \right) \tag{95}$$

where $C_1$ and $C_2$ are two adjustable model parameters. Thus, for a zero agitation rate, the above equation reduces to Eq. (92) (i.e., $\beta^{OP}_{ij} = \beta_{ij}$). On the other hand, for large values of the ratio $(J_O/J_P)$, Eq. (95) is simplified to $\beta^{OP}_{ij} = C_1 J_O^{C_2}/W_{ij}$. Coagulation rate equations similar to the last expression have often been applied to orthokinetic particle aggregation in a potential field [150].

The mean value of the shear rate $\dot{\gamma}$ within a VCM monomer droplet of radius, $R_d$, can be approximated by the well-known solution for laminar flow (Levich [151]):

$$\dot{\gamma} = \frac{\pi}{8R_d} \frac{\Delta v}{(1 + \mu_1/\mu_2)} \tag{96}$$

where $\mu_1$ and $\mu_2$ are the viscosities of the aqueous and monomer phases, respectively. $\Delta v$ is the relative droplet velocity with respect to the surrounding fluid, assuming no resistance to shear-transmission across the interface. In the open literature, several correlations have been proposed to express $\Delta v$ in terms of the average dissipation rate of turbulent kinetic energy per unit mass, $\bar{\varepsilon}$ [152].

### Effect of Primary Stabilizer

Primary stabilizers are adsorbed onto the monomer–water interface, imparting a mechanical resistance to shear-rate induced aggregation of primary particles. When an interfacial skin (i.e., a pericellular membrane) is formed around a polymerizing monomer droplet, the stirring intensity within the droplet decreases. Thus, the primary stabilizers modulate the shear-rate induced aggregation process within the droplet, effectively producing a "shielding" effect of the monomer phase to the external turbulent flow filed. An empirical approach was used to describe the reduction in average shear rate, calculated from Eq. (96) in terms of the primary stabilizer concentration $C_{PS}$:

$$\dot{\gamma}_{eff} = \dot{\gamma}\, \frac{1 + C_{PS}/C_{PS0}}{1 + C_{\dot{\gamma}}\, C_{PS}/C_{PS0}} \tag{97}$$

where $\dot{\gamma}_{eff}$ is an effective shear rate and $C_{PS0}$ is a term used to scale the primary stabilizer concentrations ($C_{PS0} = 1$ in this work). The maximum extent of shear-rate reduction is determined by the parameter $C_{\dot{\gamma}}$. Thus, Eq. (97) predicts that when $C_{PS} \ll C_{PS0}$ a zero-reduction in the shear-rate is obtained (i.e., $\dot{\gamma}_{eff} = \dot{\gamma}$). On the other hand, when $C_{PS} \gg C_{PS0}$ the maximum shear-rate reduction is obtained (i.e., $\dot{\gamma}_{eff} = \dot{\gamma}/C_{\dot{\gamma}}$).

### Effect of Secondary Stabilizer

Secondary stabilizers are assumed to partition entirely in the VCM droplets and adsorb onto the surface of the primary particles. In electrosteric stabilization of colloidal particles, the steric interaction potential is incorporated into the total electrostatic interaction potential [153, 154]. However, in the present study, due to the lack of sufficient knowledge on the exact steric stabilization mechanism of primary particles and the scarcity of experimental data, an empirical correlation was used to calculate the effective stability ratio, $W_{ij}^{eff}$, in terms of the secondary stabilizer concentration, $C_{SS}$:

$$W_{ij}^{eff}/W_{ij} = 1 + (W_S - 1)[1 - \exp(-C_{SS}/C_{SS0})] \tag{98}$$

where $C_{SS0}$ is the scaling term for $C_{SS}$ ($C_{SS0} = 1$ in this work). $W_S$ is a model parameter that controls the extent of stabilization of the primary particles by the secondary stabilizer. Thus, at very low secondary stabilizer concentrations (i.e., $C_{SS} \ll C_{SS0}$), the effective stability ratio calculated from Eq. (14) is $W_{ij}^{eff} = W_{ij}$.

On the other hand, at high secondary stabilizer concentrations (i.e., $C_{SS} \gg C_{SS0}$), the effective stability ratio is given by $W_{ij}^{\text{eff}} = W_{ij}W_S$.

### 4.1.4  Numerical Solution of the Population Balance Equation

In a series of papers on the numerical solution of the general PBEs it was shown that, for aggregation-dominated processes, the discretized population balance approach provided fast and accurate solutions to Eq. (85) [111–114]. Thus, in the present study, the discretized PBE method was used to calculate the dynamic evolution of the PPSD [101, 155].

## 4.2  Results and Discussion on Primary Particle Size Distribution

Detailed numerical simulations were carried out for a number of different VCM bulk and suspension polymerization cases. The key model variables were the polymerization temperature, $T$; the agitation rate, $N$; the initiator concentration, $C_I$; the Debye length, $\kappa^{-1}$ (or the electrolyte concentration $C_{z,i}$); the primary stabilizer concentration, $C_{PS}$; and the secondary stabilizer concentration, $C_{SS}$. In all simulations, the values of the various model parameters were kept constant, that is, $p = 1.25, D_0 = 20\,\text{nm}, C_1 = 0.01, C_2 = 1, C_{\dot{\gamma}} = 10$, and $W_S = 10$. The numerical values of $ne$ and $q$ were 40 and 2, respectively, except where otherwise noted.

The various mean particle diameters, $D_{pq}$, were calculated using the following equation:

$$D_{pq} = \left[ \frac{\sum\limits_{i=1}^{ne} D^p N_i}{\sum\limits_{i=1}^{ne} D^q N_i} \right]^{1/(p-q)} \tag{99}$$

where $p$ and $q < p$ are integers. Thus, for $p = 1$ and $q = 0$, the mean particle diameter, $D_{10}$, is obtained. Similarly, the volumetric mean particle diameter $D_{30}$ is defined for $p = 3$ and $q = 0$.

From the calculated values of the particle number distribution, $N_i(t)$, the fractional particle number distribution, $f_i(t)$, can be easily calculated:

$$f_i(t) = \frac{N_i(t)}{\sum\limits_{i=1}^{ne} N_i(t)} \tag{100}$$

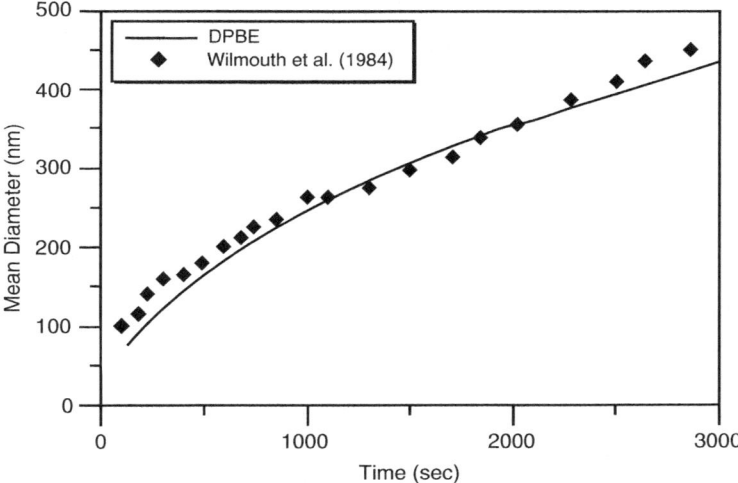

**Fig. 17** Comparison of calculated mean basic particle diameter [26] with the experimental data of Willmouth et al. [134] at 35°C and an initiator concentration of 1 g/kg VCM

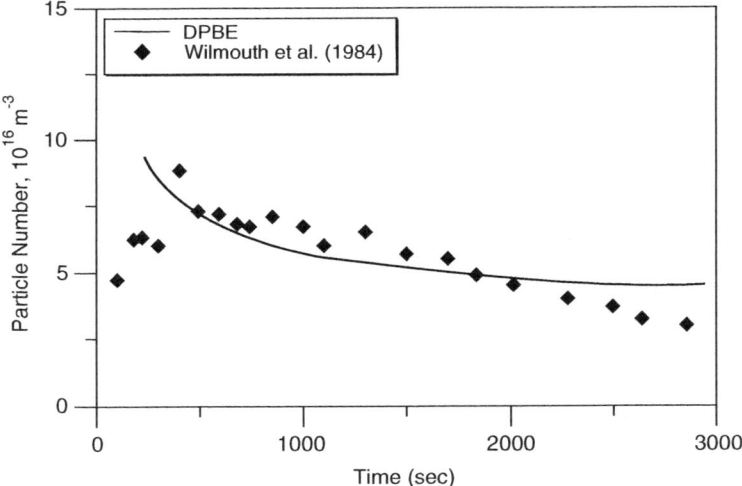

**Fig. 18** Comparison of calculated total particle number [26] with the experimental data of Willmouth et al. [134] at 35°C and an initiator concentration of 1 g/kg VCM

In Fig. 17, the experimental and calculated number average particle diameters are depicted with respect to the polymerization time at a temperature of 35°C [134]. There is very good agreement between experimental and model results. In Fig. 18, the calculated total number of particles is compared with the experimental measurements reported by Willmouth et al. [134]. Note that model predictions are

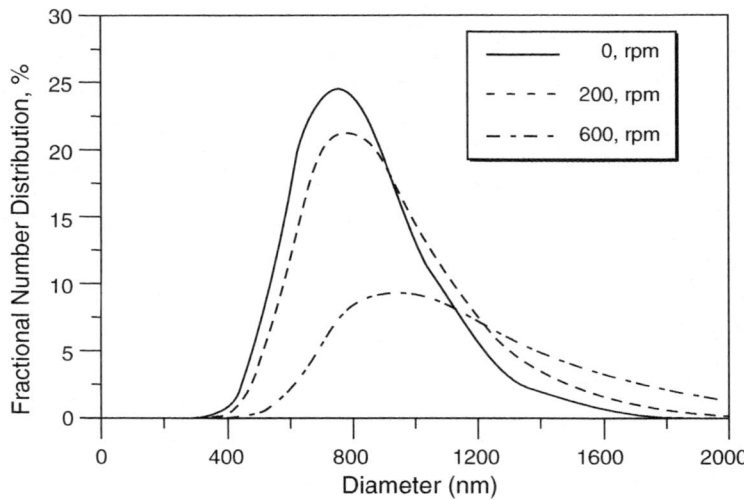

**Fig. 19** Effect of agitation rate on the mean primary particle diameter; model predictions are $C_1 = 0.01$ and $C_2 = 1$ [26]

in good agreement with experimental measurements. The observed differences between numerical simulations and experimental results, for short simulation times, can be partially attributed to limitations of the experimental technique in resolving the size of very small primary particle nuclei, as the authors themselves noted.

Tornell and Uustalu [135] performed several bulk VCM polymerization experiments in an agitated vessel in the presence of a secondary stabilizer. The polymerizations were carried out at 59°C and an agitation rate of 300 rpm in the presence of 0.2% w/w secondary stabilizer concentration. According to Tornell and Uustalu [135], massive aggregation of primary particles takes place at a monomer conversion of 7%, leading to formation of a 3D polymer skeleton.

Figures 19, 20, and 21 depict the effects of agitation rate, secondary stabilizer concentration, and electrolyte concentration on the PPSD. In these simulations, the "base case" corresponds to a non-agitated bulk VCM polymerization at 50°C and an initiator concentration (lauryl peroxide) of 1 g/kg VCM. In Fig. 19, the calculated fractional particle number distributions are plotted for different values of the agitation rate. An increase in the agitation rate results in broader distributions caused by the increased shear-induced aggregation. It is apparent that the primary particles undergo massive aggregation at 600 rpm, leading to a very broad distribution.

In Fig. 20, the calculated PPSDs at 10% VCM conversion are shown for different concentrations of secondary stabilizer. The experimental conditions correspond to the base case with $C_{SS0} = 1$ g/kg and $W_S = 10.$ An increase in secondary stabilizer concentration improves the stability of the primary particles, leading to a decrease in particle aggregation rate. As a result, the primary particle number increases while

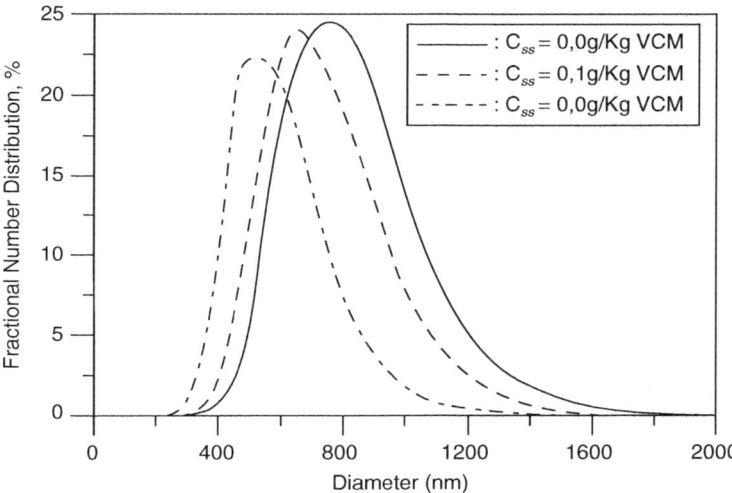

**Fig. 20** Effect of secondary stabilizer concentration on the primary particle size distribution; model predictions are $C_{SS0} = 1$ g/kg VCM and $W_S = 10$ [26]

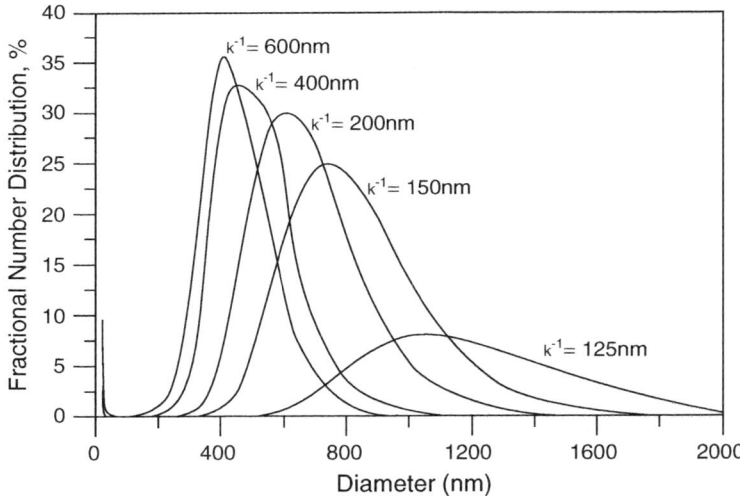

**Fig. 21** Effect of the Debye length on the PPSD (model predictions [26])

the average particle diameter decreases, in agreement with the experimental observations of Tornell and Uustalu [135].

The effect of electrolyte concentration on the PPSD was investigated in terms of the Debye length, $\kappa^{-1}$ [26, 137]. In Fig. 21, the calculated PPSDs at $t = 900$ s are shown for different values of $\kappa^{-1}$. As the value of $\kappa^{-1}$ increases (i.e., the electrolyte

**Fig. 22** Effect of
polymerization temperature
on the critical monomer
conversion for 0 and
200 rpm (model predictions
[26])

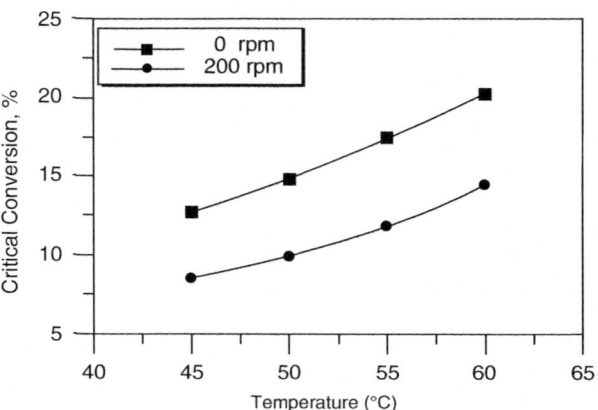

concentration decreases), the particle stability increases and the particle size decreases. Note that for $\kappa^{-1} = 125$ nm (i.e., corresponding to a high electrolyte concentration), the distribution undergoes extensive aggregation at very low monomer conversions, as indicated by the broadness of the distribution.

The effect of agitation rate on the critical monomer conversion was investigated by Davidson and Witenhafer [127] and Boissel and Fischer [141] for bulk polymerization of VCM. In Fig. 22, the variation of $X_c$ with agitation rate is shown for both bulk and suspension polymerization. The suspension polymerization results were obtained for a monomer to water ratio of 1:1 for two different primary stabilizer concentrations. As the agitation rate increases, the critical monomer conversion decreases. It should be pointed out that the decrease in critical monomer conversion with agitation rate leads to an increase in porosity, which is in qualitative agreement with the experimental observations of Smallwood [126].

## 4.3 PVC Grain Porosity

The porosity of a PVC grain is defined as the ratio of the total interior pore volume to the total volume of the grain. It is a measure of the internal structure of PVC grains and is one of the most important properties of bulk/suspension PVC resins. The porosity of a PVC grain is a consequence of the heterophase nature of bulk or suspension polymerization of VCM and is caused by multistage agglomeration of the primary particles formed directly during polymerization.

Average porosity values for PVC grains have been reported in the literature [156]. One must be careful in interpreting porosity measurements because experimental errors can be significant and the intra- and intercellular porosities are not easily distinguished from average porosity values. Density or/and pore size distribution measurements can be employed to distinguish between intra- and intercellular porosities. The contraction degree of the primary particle network within the polymerizing monomer droplets and the critical conversion of primary particle network formation are crucial factors affecting the final grain porosity. It has been shown experimentally that the final grain porosity depends on the critical monomer conversion, agitation rate, polymerization temperature, and the type and concentration of the protective colloidal stabilizers.

The driving force for the development of particle porosity is the density difference between monomer and polymer. Note that this density difference is quite large. For example, at 60°C the monomer and polymer densities are $\rho_m = 0.85$ and $\rho_p = 1.4$ g/cm$^3$, respectively. Assuming zero droplet/particle shrinkage during polymerization, the theoretical maximum value of porosity can be determined in terms of monomer and polymer densities using the following equation:

$$\varepsilon_{\max}^{th}(x) = \left(1 - X\frac{\rho_m}{\rho_p}\right) \underset{x \to 1}{\longrightarrow} \left(1 - \frac{\rho_m}{\rho_p}\right) \tag{101}$$

Thus, at a final monomer conversion $(X)$ of 100%, Eq. (101) results in a maximum value for $\varepsilon$ of 0.39 at 60°C. On the other hand, if there was no resistance to droplet/particle shrinkage, the final particle size would be 60% of its initial size and the particle porosity would be given by the following equation:

$$\varepsilon_{\min}^{th}(x) = \frac{(1 - X)}{1 - X + X(\rho_m/\rho_p)} \tag{102}$$

Thus, at a final monomer conversion of 100%, Eq. (102) results in a minimum value for $\varepsilon$ of zero.

Because the final PVC grain porosity has a finite value, some resistance to droplet/particle shrinkage is present. The resistance to droplet/particle shrinkage is generally attributed to the formation of the primary particle network within the droplet. During VCM polymerization, the primary particles grow in size, become unstable, and form primary particle aggregates. Eventually, primary particles and primary particle aggregates form a continuous 3D network inside the polymerizing monomer droplets. The structure of this network, its strength, and the conversion at which it is established, depend on the primary particle aggregation mechanism, interaction between the particles, particle size, and particle concentration. Assuming that the 3D network is formed at a critical monomer conversion of $X_c$ and that there is zero droplet/particle shrinkage from that point on, then the maximum particle porosity, $\varepsilon_{\max}(X)$, is a function of the monomer conversion and is given by the following equation:

$$\varepsilon_{\max}(X) = \left(1 - \frac{X(\rho_m/\rho_p)}{(1 - X_c + X_c\rho_m/\rho_p)}\right); \quad \text{for } X > X_c \tag{103}$$

From Eq. (103) we can easily calculate the final value of the grain porosity for $X_c$ values of 0.1, 0.2, 0.3, and 0.5, assuming a final monomer conversion of 100% and a density ratio of 0.60. The calculated values of $\varepsilon_{max}(X)$ are 0.375, 0.348, 0.318, and 0.25, respectively. Therefore, when the network formation is delayed ($X_c$ increases), the final product porosity decreases. The final porosity of the product is thought to be strongly related to the conversion at which the primary particle network is formed [126]. At critical conversion, the structure and strength of the 3D polymer skeleton depends on the size and number of primary particles and the interactions between them. The characteristics of the primary particles are, in turn, influenced by the polymerization conditions (polymerization temperature, stirrer speed, type and concentration of stabilizers, etc.)

In the case of limited contraction of the primary particle 3D network caused by particle fusion, $\varepsilon_{max}(X)$ is given by the following equation:

$$\varepsilon_{max}(x) = \left(1 - \frac{X(\rho_m/\rho_p)}{f(X)(1 - X_c + X_c\rho_m/\rho_p)}\right); \quad \text{for } X > X_c \qquad (104)$$

where $f(x)$ is the ratio of the droplet/particle volume at conversion $X > X_c$ to the droplet/particle volume at $X_c$. Note that the value of critical conversion $X_c$ can be influenced by the agitation rate [141], secondary stabilizers [135], primary stabilizers [157], VCM soluble electrolytes [136], etc. Note that in quiescent VCM polymerization, formation of a 3D polymer skeleton is substantially delayed [127] and, thus, the end-product porosity decreases significantly [141].

The effect of the critical monomer conversion on the final grain porosity is shown in Fig. 23 for a polymerization temperature of 50°C. The maximum porosity (Eq. (103) for zero droplet/particle shrinkage) and actual porosity (Eq. (104) for limited droplet/particle shrinkage) decrease as the critical monomer conversion increases. Thus, by controlling the value of $X_c$, the final value of the PVC grain porosity can be affected. The effect of monomer conversion on porosity exhibits an almost linear decrease (see Eqs. (103) and (104)). Thus, as the VCM conversion increases, droplet contraction, primary particle growth, and primary particle fusion result in a decrease in porosity. For example, the grain porosity from $\varepsilon = 0.4$ at 40% conversion decreases down to $\varepsilon < 0.1$ at $X > 70\%$. Even if the porosity seems to be decreasing toward a zero value, the polymerization usually ends at conversions of 85–95%, which means that a finite, but small, porosity is obtained.

Note that as the polymerization temperature increases, the colloidal stability of the primary particles decreases. As a result, the primary particle aggregation rate increases. This means that formation of the primary particle 3D polymer skeleton occurs at lower values of $X_c$. Thus, for zero droplet/particle shrinkage, the maximum grain porosity increases (see red line in Fig. 24). However, the decrease in critical monomer conversion is counterbalanced by an increase in primary particle fusion, which results in significant contraction (shrinkage) of the 3D polymer network and subsequent decrease in the final grain porosity (Fig. 24). Thus, fusion of the close-contact primary particles is a process that results in reduction of the

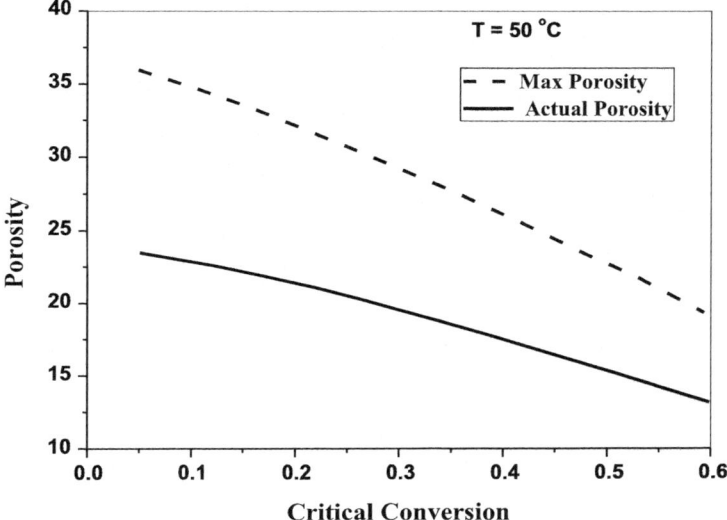

**Fig. 23** Calculated maximum and actual PVC grain porosities in terms of the critical monomer conversion at 50°C

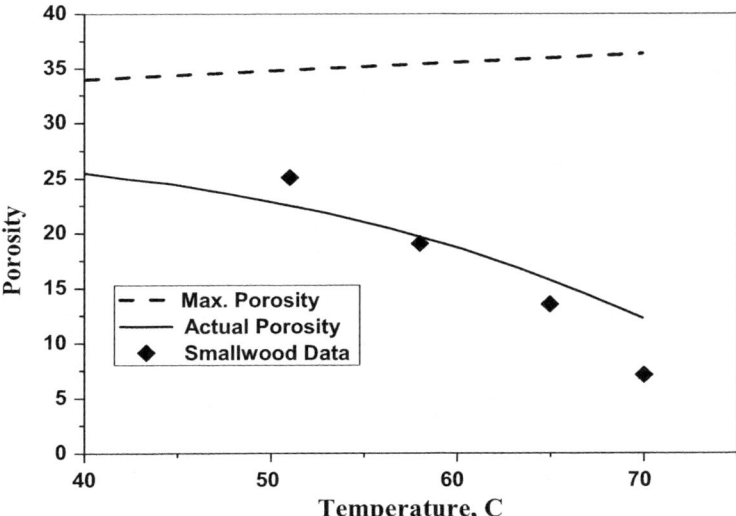

**Fig. 24** Comparison of calculated and experimental values [126] of PVC grain porosity as a function of polymerization temperature

final porosity via 3D network contraction. This primary particle fusion is caused by interfacial forces acting on the monomer-swollen PVC particles at temperatures below the glass transition temperature $(T_g)$. At 70°C, monomer-swollen PVC particles with about 23% vinyl chloride exhibit a very low $T_g$ value (about $-70°C$). The effect of temperature on particle fusion and porosity is clearly seen in Fig. 24, which shows that the porosity drops from $\varepsilon = 26\%$ to about 12% as the temperature increases from 40 to 70°C. Note also the sharp decrease in porosity at higher temperatures. This behavior is very similar to the porosity measurements of Smallwood [126], which also indicate an increased porosity loss near 70°C.

In VCM bulk polymerization, the primary particles evolve over a size range of 20 nm to 2 μm. In the early stages of polymerization, the aggregation of small primary nuclei (basic particles) is governed by a diffusive (perikinetic) mechanism. At later stages, for larger particles, shear rate induced aggregation (orthokinetic) may become important. In a quiescent VCM bulk polymerization (absence of shear rate), the primary particles continue to grow as single particles due to the long-range stabilizing electrostatic forces. As a result, they can be packed closely together without aggregating. The final product has low porosity. In the presence of agitation, the primary particles can undergo shear-induced aggregation. Thus, as the size of the primary particles increases with monomer conversion, the stirring-induced shear rate overcomes the electrostatic and steric repulsive forces of the primary particles, resulting in particle aggregation and formation of a continuous open 3D network of high porosity. By increasing the agitation rate, formation of the 3D polymer skeleton occurs at lower conversions and produces irregular grains of larger porosity [127].

In suspension polymerization, the effect of agitation rate is more complicated because of the presence of the monomer–water interface. When a pericellular membrane is formed around the polymerizing monomer droplets, external shear can be transmitted only through the membrane pores or by fluctuations of the membrane itself. Overall, the effect of agitation on shear-induced aggregation is weaker in suspension polymerization.

In suspension polymerization, it has been observed that grain porosity increases as the viscosity of the continuous aqueous phase increases. The effect of viscosity on grain porosity has been reported by Cebollada et al. [54] and Smallwood [126]. Cebollada et al. [54] studied the effect of HPMC concentration and degree of substitution on the viscosity of the continuous phase. They carried out a series of suspension polymerization experiments by varying the viscosity of the continuous phase from 0.545 to 0.992 cSt. They found that the average size and morphology of PVC particles was strongly affected by the viscosity (see Table 7). It is well known that the chemical structure of the dispersant molecules determines the physical characteristics of the adsorbed polymer film at the water–monomer interface, as well as the molecular and morphological characteristics of the semicellular membrane or skin formed on the surface of the polymerizing monomer droplets. Crosslinking of the adsorbed cellulose ethers at the monomer surface can contribute to the stability of the dispersed monomer droplets [157, 158].

Tornell and Uusatalu [136] investigated the effect of electrolyte concentration on PVC grain porosity. Because stabilization of the primary particles is achieved by the presence of negative charges on the particle surface, addition of electrolytes to the polymerization medium should decrease the stability of the particles. Thus,

**Table 7**  Effect of viscosity on grain size and grain morphology [54]

| Run | HPMC | Concentration (g/L $H_2O$) | Viscosity $v$ (cSt) | Mean pore diameter ($\mu$m) | Mean grain size ($\mu$m) | Porosity ($cm^3$/g) |
|-----|------|------|------|------|------|------|
| 1 | E15 | 0.440 | 0.545 | 0.59 | 385 | 0.179 |
| 2 | E15 | 0.644 | 0.556 | – | 181 | – |
| 3 | E15 | 0.854 | 0.568 | – | 167 | – |
| 4 | E15 | 1.343 | 0.593 | 0.54 | 117 | 0.159 |
| 5 | E50 | 0.882 | 0.593 | 0.21 | 125 | 0.086 |
| 6 | E15 | 3.000 | 0.708 | 0.23 | 80 | 0.130 |
| 7 | E50 | 2.040 | 0.708 | 0.12 | 90 | 0.054 |
| 8 | E50 | 2.990 | 0.820 | 0.16 | 105 | 0.042 |
| 9 | E50 | 4.210 | 0.992 | 0.11 | 144 | 0.095 |

**Table 8**  Effect of electrolyte ($Bu_4NBF_4$) concentration on size and porosity [135]

| Concentration (%) | Grain size | | $K$-value | Porosity (%) | Bulk density (kg/$m^3$) |
|------|------|------|------|------|------|
| | Average ($\mu$m) | Standard deviation | | | |
| 0.00 | 120 | 1.25 | 67.4 | 21.4 | 538 |
| 0.02 | 135 | 1.40 | 68.0 | 23.5 | 550 |
| 0.20 | 150 | 1.43 | 68.0 | 24.7 | 532 |

early aggregation of primary particles can lead to higher porosities, as shown in Table 8. The increase in grain size is mostly caused by the increased porosity.

Allsopp [159] reported experimental results on the effect of injected monomer before and after the reactor pressure drop at about 70% conversion. Generally, VCM polymerization at conversions close to that corresponding to the pressure drop is characterized by small amounts of monomer located in the grain pores. Because the initiator concentration in these monomer locations is relatively large, local polymerization increases and the small pores become filled with polymer. Allsopp [159] found that by injecting some monomer at conversions near the reactor pressure drop, the reaction rate and grain porosity were increased. Although the porosity characteristics were generally improved, the plasticizer adsorption was decreased. Apparently, the injected monomer reacts at pore entry regions and blocks pores. Addition of monomer before the pressure drop kept the grains from collapsing and retained their spherical shape.

The main effect of primary stabilizers is to decrease interfacial tension and, therefore, the contractile force acting on the droplets, which results in higher grain porosity. Primary stabilizers are also important in formation of the pericellular membrane. Because primary particles are sensitive to shear, it should reduce primary particle aggregation and decrease porosity. Primary stabilizers are considered to act mostly on the monomer–water interface. Secondary stabilizers are used to mediate the effects of primary stabilizers. Secondary stabilizers enter the monomer phase and their main mode of action is steric stabilization of primary particles. Therefore, secondary stabilizers affect primary particle number and size [131].

As shown in Table 7, lower porosities are obtained with HPMC E50 than with E15; both have identical substitutions but E50 has a larger molecular weight. A similar trend was reported by Cheng and Landsam [160]. A decrease in the molecular weight of HPMC resulted in an increase in grain porosity. Porosity also increased with the degree of methoxy substitution. The HPMC data indicate a steric stabilization mechanism that was more effective for the high molecular weight stabilizers.

Nilsson et al. [157] investigated the effect of primary stabilizers such as Methocel F50 (HPMC), Gohsenol GH20 (PVA with 88% DH), and Rhodoviol 5/270 (PVA with 71.5% DH) on grain porosity. The highest grain porosity was obtained with Rhodoviol 5/270 and the lowest with GH20, which was the least surface active. This clearly shows the effect of PVA DH on grain porosity. Indeed, it was found that the grain porosity decreased linearly with increasing PVA DH (e.g., from about $\varepsilon = 39\%$ at 65% DH to $\varepsilon = 28\%$ at 82.5% DH). The observed decrease in grain porosity with increase in PVA DH can be explained by the decrease in the soluble amount of PVA in the monomer phase and the resulting decrease in primary particle stabilization.

The fact that increasing the amount of steric stabilizer in the monomer phase results in increased porosity indicates that the observed porosity increase cannot be explained by delayed formation of the 3D primary particle network but by decreased fusion of primary particles as a result of decreased interfacial tension or/and by increased mechanical strength of the 3D network due to the increased number of particles and number of particle contacts.

## List of Symbols

| | |
|---|---|
| $A_0$ | Reactor outside heat transfer area, $m^2$ |
| $A_a$ | Reactor heat transfer area to the environment, $m^2$ |
| $A_i$ | Reactor inside heat transfer area, $m^2$ |
| $\alpha_m$ | Monomer activity |
| $A_t$ | Reactor top heat transfer area, $m^2$ |
| $B_i$ | Virial coefficient of $i$ component, $m^3/kmol$ |
| $B_m$ | Virial coefficient of monomer, $m^3/kmol$ |
| $B_{mw}$ | Virial coefficient of mixture monomer and water, $m^3/kmol$ |
| $B_w$ | Virial coefficient of water, $m^3/kmol$ |
| $C_1, C_2$ | Model parameters |
| $C_{pm}$ | Metal wall heat capacity, $kJ/(kg\ K)$ |
| $C_{pmix}$ | Mixture heat capacity, $kJ/(kg\ K)$ |
| $C_{PVA}$ | Concentration of the stabilizer, $kg/m^3$ |
| $C_{pw}$ | Water heat capacity, $kJ/(kg\ K)$ |
| $D_{eq}$ | Jacket equivalent diameter, m |
| $\Delta H_r$ | Specific reaction enthalpy, $kJ/kmol$ |
| $D_{imp}$ | Impeller diameter, m |

| | |
|---|---|
| $D_{pq}$ | Mean particle diameter, m |
| $D_R$ | Reactor inside diameter, m |
| $\widehat{f}$ | Fugacity, Pa |
| $f_i$ | Fractional particle number distribution |
| $f_{i,j}$ | Efficiency of initiator $i$ in the $j$ phase |
| $F_w$ | Mass flow rate of the water added to the reaction mixture, kg/h |
| $G$ | Particle growth rate due to polymerization in the polymer-rich phase, kg/s |
| $g(u)$ | Breakage rate of drops of volume $u$, $s^{-1}$ |
| $h_i$ | Heat transfer coefficient of the reaction mixture side, kJ/(m s K) |
| $h_o$ | Heat transfer coefficient from the reactor wall to jacket, kJ/(m s K) |
| $I$ | Initiator molecule |
| $I_0$ | Initial initiator concentration, g/kg VCM |
| $K$ | Solubility constant for the VCM in the aqueous phase, kg VCM/kg $H_2O$ |
| $k$ | Thermal conductivity, kW/K |
| $k(v, u)$ | Coalescence rate between two drops of volume $v$ and $u$, $m^3/s$ |
| $k_B$ | Boltzmann's constant, $m^2$ kg/($s^2$ K) |
| $k_{bj}$ | Intramolecular transfer rate constant in the $j$ phase, $s^{-1}$ |
| $k_{di,j}$ | Decomposition rate constant of initiator $i$ in the $j$ phase, $s^{-1}$ |
| $k_{fmj}$ | Chain transfer to monomer rate constant in the $j$ phase, $m^3/(kmol\ s)$ |
| $k_{fpj}$ | Chain transfer to polymer rate constant in the $j$ phase, $m^3/(kmol\ s)$ |
| $k_I$ | Rate constant for initiator decomposition, $m^3/(kmol\ s)$ |
| $k_{p1}$ | Propagation rate constant in the monomer-rich phase, $m^3/(kmol\ min)$ |
| $k_{p2}$ | Diffusion-controlled propagation rate constant in the polymer-rich phase, $m^3/(kmol\ min)$ |
| $k_{pj}$ | Propagation rate constant in the $j$ phase, $m^3/(kmol\ s)$ |
| $k_t$ | Termination rate constant in the monomer phase, $m^3/(kmol\ s)$ |
| $k_{t2}$ | Diffusion-controlled termination rate constant in the polymer-rich phase, $m^3/(kmol\ min)$ |
| $k_{t20}$ | Termination rate constant in the polymer-rich phase, $m^3/(kmol\ min)$ |
| $k_{tcj}$ | Termination by combination rate constant in the $j$ phase, $m^3/(kmol\ s)$ |
| $k_{tdj}$ | Termination by disproportionation rate constant in the $j$ phase, $m^3/(kmol\ s)$ |
| $k_{zj}$ | Inhibition rate constant in the $j$ phase, $m^3/(kmol\ s)$ |
| $L_{eq}$ | Jacket equivalent length, m |
| $L_n$ | Number of long chain branches per polymer molecule |
| $M$ | Mass of monomer, kg |
| $M_0$ | Initial mass of monomer, kg |
| $M_n$ | Number average molecular weight, kg/kmol |
| $M_w$ | Weight average molecular weight, kg/kmol |
| $MW_m$ | Molecular weight of monomer, kg/kmol |
| $MW_w$ | Molecular weight of water, kg/kmol |
| $MW_x$ | Molecular weight of molecular species "$x$", kg/kmol |
| $N$ | Agitation rate, rpm |
| $n(v, t)$ | Number density function, $m^{-6}$ |

| | |
|---|---|
| $n_0(v)$ | Initial drop size distribution of the dispersed phase, $m^{-6}$ |
| $N_d$ | Number of initiators used in the polymerization |
| $N_{da}$ | Number of daughter drops per breakage event |
| $N_{sa}$ | Number of satellite drops per breakage event |
| $N_i$ | Particle number distribution |
| $N_{We}$ | Weber number |
| $P$ | Total reactor pressure, Pa |
| $P_m$ | Monomer partial pressure, Pa |
| $P_m{}^{sat}$ | Monomer saturation pressure, Pa |
| $Pr$ | Prandtl number |
| $P_w{}^{sat}$ | Water saturation pressure, Pa |
| $[P_x]$ | "Dead" polymer chains, containing $x$ monomer units, $kmol/m^3$ |
| $R$ | Ideal gas constant, J/mol/K |
| $r$ | Radius of colloidal particles, m |
| $Re$ | Reynolds number |
| $R_{pm}$ | Polymerization rates in the monomer-rich phase, $kmol/(m^3 \ s)$ |
| $R_{pp}$ | Polymerization rates in the polymer-rich phase, $kmol/(m^3 \ s)$ |
| $[R_x^{\bullet}]$ | "Live" macroradicals, containing $x$ monomer units, $kmol/m^3$ |
| $r_{\lambda j,j}$ | "Live" polymer moment rate function, $kmol/(m^3 \ s)$ |
| $r_{\mu j}$ | "Dead" polymer moment rate function, $kmol/(m^3 \ s)$ |
| $S_0$ | Nucleation rate of primary particles of volume $v_0$ in the monomer-rich phase, $s^{-1}$ |
| $S_d$ | Number density of SCB per 1,000 monomer units |
| $S_n$ | Number of short chain branches per polymer molecule |
| $t$ | Time, s |
| $T$ | Reactor mixture temperature, K |
| $T_0$ | Reference temperature, K |
| $T_a$ | Ambient temperature, K |
| $T_j$ | Reactor's jacket temperature, K |
| $T_m$ | Temperature of the metal wall, K |
| $T_n$ | Number of terminal double bonds per polymer molecule |
| $u(u)$ | Number of droplets formed by the breakage of a drop of volume $u$ |
| $U_a$ | Heat transfer coefficient to the reactor environment, $kJ/(m \ s \ K)$ |
| $U_t$ | Heat transfer coefficient from the reactor top, $kJ/(m \ s \ K)$ |
| $V$ | Total volume of the polymer particles, $m^3$ |
| $v_{da}$ | Volume of daughter drops, $m^3$ |
| $v_{sa}$ | Volume of satellite drops, $m^3$ |
| $V_f$ | Free volume of the mixture in the polymer-rich phase, $m^3$ |
| $V_f{}^*$ | Free volume of the mixture at the critical monomer conversion, $m^3$ |
| $V_{fm}$ | Free volume of monomer, $m^3$ |
| $V_{fp}$ | Free volume of polymer, $m^3$ |
| $V_g$ | Volume of gas phase, $m^3$ |
| $V_j$ | Volume of $j$-phase, $m^3$ |
| $V_m$ | Metal wall volume, $m^3$ |

| | |
|---|---|
| $V_{mix}$ | Reaction mixture volume, $m^3$ |
| $V_R$ | Total reactor volume, $m^3$ |
| $W_{ij}$ | Fuch's stability ratio |
| $W_w$ | Total mass of water loaded in the reactor, kg |
| $W_{w\alpha}$ | Mass of water in the aqueous phase, kg |
| $X$ | Monomer conversion |
| $X_c$ | Critical monomer conversion |
| $X_f$ | Fractional monomer conversion |
| $y_m$ | Mole fraction of monomer in the vapor phase |
| $y_w$ | Mole fraction of water in the vapor phase |
| $Z$ | Inhibitor molecule |

## Greek Symbols

| | |
|---|---|
| $\beta$ | Aggregation rate kernel, $s^{-1}$ |
| $\beta(u, v)$ | Daughter drop breakage function, accounting for the probability that a drop of volume $v$ is formed via the breakage of a drop of volume $u$, $m^{-3}$ |
| $\dot{\gamma}$ | Mean value of the shear rate, $s^{-1}$ |
| $\dot{\gamma}_{eff}$ | Effective shear rate, $s^{-1}$ |
| $\Delta v$ | Relative droplet velocity, m/s |
| $\varepsilon$ | Porosity |
| $\bar{\varepsilon}$ | Average dissipation rate of turbulent kinetic energy per unit mass, $m^2/s^3$ |
| $\eta$ | Kinematic viscosity, $m^2/s$ |
| $[\eta]$ | Intrinsic viscosity, $m^3/kg$ |
| $\theta$ | Surface coverage |
| $\kappa^{-1}$ | Debye length, m |
| $[\lambda_{i,j}]$ | $i$-th moment of molecular weight distribution of "live" polymer radicals in the $j$ phase, $kmol/m^3$ |
| $\lambda_b$ | Breakage coalescence efficiency |
| $\lambda_c$ | Coalescence efficiency |
| $\mu$ | Viscosity, kg/(m s) |
| $[\mu_k]$ | $k$-th moment of dead polymer chains, $kmol/m^3$ |
| $\rho$ | Density, $kg/m^3$ |
| $\rho_m$ | Monomer density, $kg/m^3$ |
| $\rho_{mix}$ | Mixture density, $kg/m^3$ |
| $\rho_p$ | Polymer density, $kg/m^3$ |
| $\rho_w$ | Water density, $kg/m^3$ |
| $\sigma$ | Interfacial tension, $kg/s^2$ |
| $\sigma_0$ | Standard deviation |
| $\sigma_{da}$ | Standard deviation of the distribution for daughter drops |
| $\sigma_{sa}$ | Standard deviation of the distribution for satellite drops |
| $\varphi$ | Volume fraction of the dispersed phase |
| $\varphi_1$ | Volume fraction of the monomer-rich phase |

| | |
|---|---|
| $\varphi_2$ | Volume fraction of the polymer-rich phase |
| $\varphi_{2,C}$ | Critical value of the polymer volume fraction in the polymer-rich phase |
| $\varphi_{cr}$ | Polymer volume fraction corresponding to the critical monomer conversion |
| $\varphi_j$ | Volume fraction of the polymer in the $j$ phase |
| $\varphi_{pol}$ | Volume fraction of the polymer in the dispersed phase |
| $\widehat{\phi}_m$ | Fugacity coefficient of monomer |
| $\chi$ | Flory–Huggins interaction parameter |
| $\omega_b$ | Breakage frequency, $s^{-1}$ |
| $\omega_c$ | Collision frequency, $s^{-1}$ |

## Superscripts

| | |
|---|---|
| $g$ | Gas phase |
| $m$ | Monomer phase |
| $p$ | Polymer phase |
| $w$ | Aqueous phase |

## References

1. Smallwood PV (1990) Vinyl chloride polymers, polymerization. In: Mark H (ed) Encyclopedia of polymer science and technology, vol 17. Wiley, New York, p. 295
2. Saeki Y, Emura T (2002) Technical progresses for PVC production. Prog Polym Sci 27:2055
3. Burgess RH (1982) Manufacturing and processing of PVC. Applied Science, London
4. Langsam M (1986) In: Nass LI, Heiberger CA (eds) Encyclopedia of PVC, vol 1, 2nd edn. Marcel Dekker, New York, p. 48
5. Tornell BE (1988) Recent developments in PVC polymerization. Polym-Plast Technol Eng 27:1
6. Xie TY, Hamielec AE, Wood PE, Woods DR (1991) Suspension, bulk and emulsion polymerization of vinyl chloride–mechanism, kinetics and modelling. J Vinyl Technol 13 (1):2
7. Xie TY, Hamielec AE, Wood PE, Woods DR (1991) Experimental investigation of vinyl chloride polymerization at high conversion: mechanism, kinetics and modelling. Polymer 32 (3):537
8. Darvishi R, Esfahany MN, Bagheri R (2015) S-PVC grain morphology: a review. Ind Eng Chem Res 54(44):10953–10963
9. Mejdell T, Pettersen T, Naustdal C, Svendsen HF (1999) Modelling of industrial S-PVC reactor. Chem Eng Sci 54:2459
10. Sidiropoulou E, Kiparissides C (1990) Mathematical modelling of PVC suspension polymerization. J Makromol Sci Chem A27(3):257
11. Abdel-Alim AH, Hamielec AE (1972) Bulk polymerization of vinyl chloride. J Appl Polym Sci 16:783
12. Kuchanov SI, Bort GC (1973) Kinetics and mechanism of polymerization of vinyl chloride. Polym Sci A15:2712

13. Ray WH, Jain SK, Salovey R (1975) On the modelling of bulk PVC reactors. J Appl Polym Sci 19:1297
14. Ugelstad J, Moerk PC, Hansen FK, Kaggerund KH, Ellingsen T (1981) Kinetics and mechanism of vinyl chloride polymerization. Pure Appl Chem 53:323
15. Chan RKS, Langsam M, Hamielec AE (1982) Calculation and applications of VCM distribution in vapor/water/solid phase during VCM polymerization. J Macromol Sci Chem A17 (6):969
16. Hamielec AE, Gomez-Vaillard R, Marten FL (1982) Diffusion controlled free radical polymerization. Effect on polymerization rate and molecular properties of PVC. J Macromol Sci Chem A17(6):1005
17. Kelsall DG, Maitland GC (1983) The interaction of process conditions and product properties for PVC. Munich, Polymer Reaction Engineering. Technical University of Berlin, Berlin, pp. 131–152
18. Weickert G, Henschel G, Weissenborn KD (1987) Kinetik der VC polymerisation. Angew Makromol Chem 147:1
19. Weickert G, Henschel G, Weissenborn KD (1987) Kinetik der VC polymerisation. Angew Makromol Chem 147:19
20. Xie TY, Hamielec AE, Wood PE, Woods DR (1991) Experimental investigation of vinyl chloride polymerization at high conversion: reactor dynamics. J Appl Polym Sci 43:1259
21. Dimian A, Van Diepen D, Van der Wal GA (1995) Dynamic simulation of a PVC suspension reactor. Comput Chem Eng 19S:S427
22. Lewin DR (1996) Modelling and control of an industrial PVC suspension polymerization reactor. Comput Chem Eng 20:S865
23. Kiparissides C, Daskalakis G, Achilias DS, Sidiropoulou E (1997) Dynamic simulation of industrial poly(vinyl chloride) batch suspension polymerization reactors. Ind Eng Chem Res 36:1253
24. Krallis A, Kotoulas C, Papadopoulos S, Kiparissides C, Bousquet J, Bonardi C (2004) A comprehensive kinetic model for the free-radical polymerization of vinyl chloride in the presence of monofunctional and bifunctional initiators. Ind Eng Chem Res 43:6382
25. Kotoulas C, Kiparissides C (2006) A generalized population balance model for the prediction of particle size distribution in suspension polymerization reactors. Chem Eng Sci 61:332
26. Alexopoulos A, Kiparissides C (2007) On the prediction of internal particle morphology in suspension polymerization of vinyl chloride. Part I: the effect of primary particle size distribution. Chem Eng Sci 62:3970
27. Starnes Jr WH, Zaikov VC, Chung HT, Wojciechowski BJ, Tran HV, Saylor K (1998) Intramolecular hydrogen transfers in vinyl chloride polymerization: routes to doubly branched structures and internal double bonds. Macromolecules 31:1508
28. Starnes Jr WH (2002) Structural and mechanistic aspects of the thermal degradation of poly (vinyl chloride). Prog Polym Sci 27:2133
29. Van Cauter K, Van Den Bossche BJ, Van Speybroeck V, Waroquier M (2007) Ab initio study of free-radical polymerization: defect structures in poly(vinyl chloride). Macromolecules 40:1321–1331
30. Van Cauter K, Van Speybroeck V, Waroquier M (2007) Ab initio study of poly(vinyl chloride) propagation kinetics: head-to-head versus head-to-tail additions. Chem Phys Chem 8:541–552
31. Wieme J, Marin GB, Reyniers M-F (2007) Modelling the formation of structural defects during the suspension polymerization of vinyl chloride. Chem Eng Sci 62:5300–5303
32. Wieme J, D'hooge DR, Reyniers M-F, Marin GB (2009) Importance of radical transfer in precipitation polymerization: the case of vinyl chloride suspension polymerization. Macromol React Eng 3:16–35
33. De Roo T, Wieme J, Heynderickx GJ, Marin GB (2005) Estimation of intrinsic rate coefficients in vinyl chloride suspension polymerization. Polymer 46:8340–8354

34. Dos Santos FN, Horiuchib LN, Pereira PAP (2014) Development of a method for the identification of organic contaminants in vinyl chloride monomer (VCM) by TD-GC-MS and multivariate analysis. Anal Methods 6:8946–8955
35. Achilias D, Kiparissides C (1992) Toward the development of a general framework for modeling molecular weight and compositional changes in free radical copolymerization reactions. J Macromol Sci Part C: Polym Rev C32:183–234
36. Mastan E, Zhu S (2015) Method of moments: a versatile tool for deterministic modeling of polymerization kinetics. Eur Polym J 68:139–160
37. D'hooge DR, Van Steenbergea PHM, Reyniersa MF, Marin GB (2016) The strength of multi-scale modeling to unveil the complexity of radical polymerization. Prog Polym Sci 58:59–89
38. Baltsas A, Achilias D, Kiparissides C (1996) A theoretical investigation of the production of branched copolymers in continuous stirred tank reactors. Macromol Theory Simul 5:477–497
39. Xie TY, Hamielec AE, Wood PE, Woods DR (1987) Experimental investigation of vinyl chloride polymerization at high conversion-temperature/pressure/conversion and monomer phase distribution relationships. J Appl Polym Sci 34:1749–1766
40. Achilias D, Kiparissides C (1992) Development of a general mathematical framework for modelling diffusion-controlled free radical polymerization reactions. Macromolecules 25:3739–3750
41. De Roo T, Heynderickx GJ, Marin GB (2004) Diffusion-controlled reactions in vinyl chloride suspension polymerization. Macromol Symp 206:215–228
42. Wieme J, Reyniers M-F, Marin GB (2009) Initiator efficiency modeling for vinyl chloride suspension polymerization. Chem Eng J 154:203–210
43. Allsopp MW (1982) In: Burgess RH (ed) Manufacture and processing of PVC. Applied Science Publishers, London
44. Allsopp MW, Vianello G (1992) Poly(vinyl chloride). In: Ullmann's encyclopedia of industrial chemistry, vol A21. Wiley-VCH, New York, pp. 717–742
45. Yuan HG, Kalfas G, Ray WH (1991) Suspension polymerization. J Macromol Sci Part C: Polym Rev C31:215–299
46. Brooks BW (2010) Suspension polymerization processes. Chem Eng Technol 33 (11):1737–1744
47. Bárkányi A, Németh S, Lakatos BG (2013) Modelling and simulation of suspension polymerization of vinyl chloride via population balance model. Comput Chem Eng 59:211–218
48. Chatzi EG, Kiparissides C (1992) Dynamic simulation of bimodal drop size distributions in low coalescence batch dispersion systems. Chem Eng Sci 47:445–456
49. Chatzi EG, Boutris CJ, Kiparissides C (1991) On-line monitoring of drop size distributions in agitated vessels. 1. Effects of temperature and impeller speed. Ind Eng Chem Res 30:536–543
50. Hartland S (1968) The coalescence of a liquid drop at a liquid-liquid interface. Part V: the effect of surface active agent. Trans Inst Chem Eng (London) 46:T275
51. Hamielec AE, Tobita H (1992) Polymerization processes. In: Ullmann's encyclopedia of industrial chemistry, vol A21. Wiley-VCH, New York, pp. 305–428
52. Chatzi EG, Kiparissides C (1994) Drop size distributions in high holdup fraction suspension polymerization reactors: effect of the degree of hydrolysis of PVA stabilizer. Chem Eng Sci 49:5039–5052
53. Cheng JT, Langsam M (1985) Particle structure of PVC based on cellulosic suspension system. III. Effect of monomer refluxing. J Appl Polym Sci 30:1365–1378
54. Cebollada AF, Schmidt MJ, Farber JN, Cariati NJ, Valles EM (1989) Suspension polymerization of vinyl chloride. I. Influence of viscosity of suspension medium on resin properties. J Appl Polym Sci 37:145–166
55. Chatzi EG, Kiparissides C (1995) Steady state drop size distribution in high holdup fraction dispersion systems. AICHE J 41:1640–1652
56. Nilsson H, Silvegren C, Tornell B (1985) Suspension stabilizers for PVC production. I. Interfacial tension measurements. J Vinyl Technol 7(3):112–118

57. Lankveld JM, Lyklema J (1972) Adsorption of polyvinyl alcohol on the paraffin-water interface: I. Interfacial tension as a function of time and concentration. J Colloid Interface Sci 41:454–462

58. Maggioris D, Goulas A, Alexopoulos AH, Chatzi EG, Kiparissides C (2000) Prediction of particle size distribution in suspension polymerization reactors: effect of turbulence nonhomogeneity. Chem Eng Sci 55:4611–4627

59. Kiparissides C, Achilias DS, Chatzi E (1994) Dynamic simulation of primary particle-size distribution in vinyl chloride polymerization. J Appl Polym Sci 54:1423–1438

60. Kumar S, Ramkrishna D (1996) On the solution of population balance equations by discretization–I. A fixed pivot technique. Chem Eng Sci 51:1311–1332

61. Shinnar R, Church JM (1960) Predicting particle size in agitated dispersions. Ind Eng Chem 52:253–256

62. Hinze JO (1955) Fundamentals of the hydrodynamic mechanism of splitting in dispersion processes. AICHE J 1:289–295

63. Narsimhan G, Gupta G, Ramkrishna D (1979) A model for translational breakage probability of droplets in agitated lean liquid-liquid dispersions. Chem Eng Sci 34:257–265

64. Ward JP, Knudsen JG (1967) Turbulent flow of unstable liquid-liquid dispersions: drop sizes velocity distributions. AICHE J 13:356–367

65. Chen HT, Middleman S (1967) Drop size distribution in agitated liquid-liquid systems. AICHE J 13:989–996

66. Doulah MS (1975) An effect of hold-up on drop sizes in liquid-liquid dispersions. Ind Eng Chem Fundam 14(2):137–138

67. Coulaloglou CA, Tavlarides LL (1977) Description of interaction processes in agitated liquid-liquid dispersions. Chem Eng Sci 32:1289–1297

68. Wang CY, Calcabrese RV (1986) Drop breakup in trurbulent stirred-tank contactors. Part II: relative influence of viscosity and interfacial tension. AICHE J 32(4):667–676

69. Calabrese RV, Chang TPK, Dang PT (1986) Drop breakup in turbulent stirred-tank contactors. Part I: effect of dispersed phase viscosity. AICHE J 32:657–666

70. Lagisetty JS, Das PK, Kumar R, Gandhi KS (1986) Breakage of viscous and non-Newtonian drops in stirred dispersions. Chem Eng Sci 41:65–71

71. Laso M, Steiner L, Hartland S (1987) Dynamic simulation of agitated liquid-liquid disperions–II experimental determination of breakage and coalescence rates in a stirred tank. Chem Eng Sci 42:2437–2446

72. Chatzi EG, Gavrielides A, Kiparissides C (1989) Generalized model for prediction of the steady-state drop size distributions in batch stirred vessels. Ind Eng Chem Res 28:1704

73. Chatzi EG, Boutris CJ, Kiparissides C (1991) On-line monitoring of drop size distribution in agitated vessels. 2. Effect of stabilizer concentration. Ind Eng Chem Res 29:1307–1316

74. Zerfa M, Brooks BW (1996) Vinyl chloride dispersion with relation to suspension polymerization. Chem Eng Sci 51(14):3591–3611

75. Vivaldo-Lima E, Wood PE, Hamielec AE (1997) An updated review on suspension polymerization. Ind Eng Chem Res 36:939–965

76. Shinnar R (1961) On the behaviour of liquid dispersions in mixing vessels. J Fluid Mech 10:259–271

77. Arai K, Konno M, Matunaga Y, Saito S (1977) Effect of dispersed-phase viscosity on the maximum stable drop size for breakup in turbulent flow. J Chem Eng Jpn 10:325–239

78. Alvarez J, Alvarez J, Hernandez M (1994) A population balance approach for the description of particle size distribution in suspension polymerization reactors. Chem Eng Sci 49:99–113

79. Tsouris C, Tavlarides LL (1994) Breakage and coalescence models for drops in turbulent dispersions. AICHE J 40(3):395–406

80. Sathyagal AN, Ramkrishna D, Narshimhan G (1996) Droplet breakage in stirred dispersions. Breakage functions from experimental drop-size distributions. Chem Eng Sci 51 (9):1377–1391

81. Chen Z, Pruss J, Warnacke H-J (1998) A population balance models for disperse systems: drop size distribution in emulsion. Chem Eng Sci 53(5):1059–1066
82. Kotoulas C, Kiparissides C (2007) Suspension polymerization. In: Asua JM (ed) Polymer reaction engineering. Blackwell, Oxford, pp. 209–230
83. Howarth WJ (1964) Coalescence of drops in a turbulent flow field. Chem Eng Sci 19:33–42
84. Delichatsios MA, Probstein RF (1976) The effect of coalescence on the average drop size in liquid-liquid dispersions. Ind Eng Chem Fundam 15:134–138
85. Sovova H (1981) Breakage and coalescence of drops in a batch stirred vessel. II. Comparison of model and experiments. Chem Eng Sci 36:1567–1573
86. Muralidhar R, Ramkrishna D (1986) Analysis of droplet coalescence in turbulent liquid-liquid dispersions. Ind Eng Chem Fundam 25:554–560
87. Muralidhar R, Ramkrishna D, Das PK, Kumar R (1988) Coalescence of rigid droplets in a stirred dispersion—II. Comparison of model and experiments. Chem Eng Sci 43:1559–1568
88. Kumar S, Kumar R, Gandhi KS (1993) A new model for coalescence efficiency of drops in stirred dispersions. Chem Eng Sci 48(11):2025–2038
89. Calabrese RV, Pacek AW, Nienow AW (1993) Coalescence of viscous drops in a stirred dispersion. In: The 1993 ICHEME research event. Institute of Chemical Engineers, London, pp. 642–645
90. Liu S, Li D (1999) Drop coalescence in turbulent dispersions. Chem Eng Sci 54:5667–5675
91. Bouyatiotis BA, Thornton JD (1967) Liquid-liquid extraction studies in stirred tanks. Part I. Droplet size and hold-up measurements in a seven-inch diameter baffled vessel. Instit Chem Eng (London) Symp Ser 26:43–51
92. Vermeulen T, Williams GM, Langlois GE (1995) Interfacial area in liquid-liquid and gas-liquid agitation. Chem Eng Prog 51:85F
93. Krieger IM (1972) Rheology of monodispersed lattices. Adv Colloid Interf Sci 3:111–136
94. Okaya T (1992) General properties of polyvinyl alcohol in relation to its applications. In: Finch CA (ed) Polyvinyl alcohol developments. Wiley, London
95. Defay R, Prigogine I, Bellemans A, Everett DH (1966) Surface tension and adsorption. Wiley, New York
96. Siow KS, Patterson D (1973) Surface thermodynamics of polymer solutions. J Phys Chem 77(3):356–368
97. Kiparissides C, Alexopoulos A, Roussos A, Dompazis G, Kotoulas C (2004) Population balance modelling of particulate polymerization processes. Ind Eng Chem Res 43:7290–7302
98. Hidy GM (1965) On the theory of the coagulation of noninteracting particles in Brownian motion. J Colloid Sci 20:123–144
99. Marchal P, David R, Klein JP, Villermaux J (1988) Crystallization and precipitation engineering-I. An efficient method for solving population balance in crystallization with agglomeration. Chem Eng Sci 43(1):59–67
100. Batterham RJ, Hall JS, Barton G (1981) Pelletizing kinetics and simulation for full-scale balling circuits. In: Proceedings 3rd International Symposium on aggregation, Nurnberg, W. Germany. A136
101. Hounslow MJ, Ryall RL, Marshall VR (1988) Discretized population balance for nucleation, growth, and aggregation. AICHE J 34(11):1821–1832
102. Kumar S, Ramkrishna D (1996) On the solution of population balance equations by discretization-II. A moving pivot technique. Chem Eng Sci 51(8):1333–1342
103. Bleck R (1970) A fast, approximate method for integrating the stochastic coalescence equation. J Geophys Res 75:5165–5171
104. Gelbard F, Seinfeld JH (1980) Simulation of multicomponent aerosol dynamics. J Colloid Interface Sci 78(2):485–501
105. Sastry KVS, Gaschignard P (1981) Discretization procedure for the coalescence equation of particulate processes. Ind Eng Chem Fundam 20:355–361
106. Gelbard F, Seinfeld JH (1979) Exact solution of the general dynamic equation for aerosol growth by condensation. J Colloid Interface Sci 68(1):173–183

107. Nicmanis M, Hounslow MJ (1998) Finite-element methods for steady-state population balance equations. AICHE J 44:2258–2272
108. Chen M-Q, Hwang C, Shih Y-P (1996) A wavelet-Galerkin method for solving population balance equations. Comput Chem Eng 20(2):131–145
109. Ramkrishna D (1985) The status of population balances. Rev Chem Eng 3(1):49–95
110. Dafniotis P (1996) Modelling of emulsion copolymerization reactors operating below the critical micelle concentration. PhD thesis, University of Wisconsin-Madison
111. Alexopoulos AH, Roussos AI, Kiparissides C (2004) Part I: dynamic evolution of the particle size distribution in particulate processes undergoing combined particle growth and aggregation. Chem Eng Sci 59:5751–5769
112. Alexopoulos AH, Kiparissides C (2005) Part II: dynamic evolution of the particle size distribution in particulate processes undergoing simultaneous particle nucleation, growth and aggregation. Chem Eng Sci 60:4157–4169
113. Roussos AI, Alexopoulos AH, Kiparissides C (2005) Part III: dynamic evolution of the particle size distribution in batch and continuous particulate processes: a Galerkin on finite elements approach. Chem Eng Sci 60:6998–7010
114. Meimaroglou D, Roussos AI, Kiparissides C (2006) Part IV: dynamic evolution of the particle size distribution in particulate processes. A comparative study between Monte Carlo and the generalized method of moments. Chem Eng Sci 61:5620–5635
115. Johnson GR (1980) Effects of agitation during VCM suspension polymerization. J Vinyl Technol 2:138–140
116. Etesami N, Nasr Esfahany M, Bagheri R (2008) Effect of the phase ratio on the particle properties of poly(vinyl chloride) resins produced by suspension polymerization. J Appl Polym Sci 110:2748–2755
117. Etesami N, Nasr Esfahany M, Bagheri R (2010) Experimental study of the effect of reflux rate during suspension polymerization on particle properties of PVC resin. Ind Eng Chem Res 49:1997–2002
118. Etesami N, Nasr Esfahany M, Bagheri R (2010) Investigation of the effect of delayed reflux on PVC grain properties produced by suspension polymerization. J Appl Polym Sci 117:2506–2514
119. Alexopoulos AH, Maggioris D, Kiparissides C (2002) CFD analysis of turbulence non-homogeneity in mixing vessels. A two-compartment model. Chem Eng Sci 57:1735–1752
120. Oldshue JY (1983) Fluid mixing technology. McGraw-Hill, New York
121. Okufi S, Perez de Ortiz ES, Sawistowski H (1990) Scale-up of liquid-liquid dispersions in stirred tanks. Can J Chem Eng 68:400–406
122. Scully DB (1976) Scale-up in suspension polymerization. J Appl Polym Sci 20:2299–2303
123. Lewis MH, Johnson GR (1981) Agitation scale-up effects during VCM suspension polymerization. J Vinyl Technol 3(2):102–106
124. Ozkaya N, Erbay E, Bilgic T, Savasci T (1993) Agitation scale-up model for suspension polymerization of vinyl chloride. Angew Makromol Chem 211:35–51
125. Tregan R, Bonnemayre A (1970) Rev Plast Mod 23:7
126. Smallwood PV (1986) The formation of grains of suspension poly(vinyl chloride). Polymer 27:1609–1618
127. Davidson JA, Witenhafer DE (1980) Particle structure of suspension poly(vinyl chloride) and its origin in the polymerization process. J Appl Polym Sci Polym Phys Ed 18:51–69
128. Nilsson H, Norvitt T, Silvegren C, Tornell B (1985) Suspension stabilizers for PVC production II: drop size distribution. J Vinyl Technol 7(3):119–122
129. Allsopp MW (1981) The development of suspension PVC morphology. Pure Appl Chem 53:449–465
130. Marquez EF, Lagos LL (2004) Mathematical modeling of the porosity of suspension poly (vinyl chloride). AICHE J 50:3184–3194

131. Tornell BE, Uustalu JM (1982) The influence of additives on the primary particle nucleation and agglomeration in poly(vinyl-chloride). J Vinyl Technol 4(2):53–56
132. Geil PH (1977) Polymer morphology. J Macromol Sci Chem A11(7):1271–1280
133. Ravey M (1977) Mechanism of scale formation in PVC reactors. J Appl Polym Sci 21:839–840
134. Willmouth FM, Rance DG, Henman KM (1984) An investigation of precipitation polymerization in liquid vinyl chloride by photon correlation spectroscopy. Polymer 25:1185–1192
135. Tornell BE, Uustalu JM (1988) Formation of primary particles in vinyl chloride polymerization. J Appl Polym Sci 35:63–74
136. Tornell BE, Uustalu JM (1986) Primary particle stability in bulk polymerization of vinyl chloride at high ion strength. Polymer 27:250–252
137. Tornell BE, Uustalu JM, Jonsson B (1986) Colloidal stability of PVC primary particles in vinyl chloride. Colloid Polym Sci 264:439–444
138. Rance DG, Zichy EL (1981) The life-cycle of the two-phase system in vinyl chloride polymerization. Pure Appl Chem 53:377–384
139. Wilson JC, Zichy EL (1979) Observations of charge on nascent poly(vinyl chloride) particles in monomer. Polymer 20(2):264–265
140. Verwey EJW, Overbeek JTG (1948) Theory of the stability of lyophobic colloids. Dover, New York
141. Boissel J, Fischer N (1977) Bulk polymerization of vinyl chloride: nucleation phase. Macromol Sci Chem A11(7):1249–1269
142. Kiparissides C, Moustakis I, Hamielec A (1993) Electrostatic and steric stabilization of PVC particles. J Appl Polym Sci 49:445–459
143. Ramkrishna D (2000) Population balances: theory and applications to particulate systems in engineering. Academic, San Diego
144. Kiparissides C (1990) Prediction of the primary particle size distribution in vinyl chloride polymerization. Macromol Chem Macromol Symp 35(36):171–192
145. Talamini G, Visentini A, Kerr J (1998) Bulk and suspension polymerization of vinyl chloride: the two-phase model. Polymer 39(10):1879–1891
146. Endo K (2002) Synthesis and structure of poly(vinyl chloride). Prog Polym Sci 27:2021–2054
147. Fuchs NA (1964) The mechanics of aerosols. Pergamon, New York
148. Van de Ven TGM (1989) Colloidal hydrodynamics. Colloid Science, vol 4. Academic, New York
149. Van de Ven TGM, Mason SG (1977) The micro-rheology of colloidal dispersions. Part VIII: effect of shear on perikinetic doublet formation. Colloid Polym Sci 255:794–804
150. Chern CS, Kuo YN (1996) Shear-induced coagulation kinetics of semibatch seeded emulsion polymerization. Chem Eng Sci 51(7):1079–1087
151. Levich V (1962) Physicochemical hydrodynamics. Academic, London
152. Batchelor GK (2000) An introduction to fluid mechanics. Cambridge University Press, Cambridge
153. Einarson MB, Berg JC (1993) Electrosteric stabilization of colloidal dispersions. J Colloid Interface Sci 155(1):165–172
154. Lazaridis N, Alexopoulos AH, Chatzi EG, Kiparissides C (1999) Steric stabilization in emulsion polymerization using oligomeric nonionic surfactants. Chem Eng Sci 54:3251–3261
155. Litster JD, Smit DJ, Hounslow MJ (1995) Adjustable discretized population balance for growth and aggregation. AICHE J 41:591–603
156. Salovey R, Cortellucci R, Roaldi A (1974) The surface area of bulk poly(vinyl chloride). Polym Eng Sci 14(2):120–123
157. Nilsson H, Silvegren C, Tornell B, Lundqvist J, Pettersson S (1985) Suspension stabilizers for PVC production III: control of resin porosity. J Vinyl Technol 7(3):123–127

158. Sarkar N, Archer WL (1991) Utilizing cellulose ethers as suspension agents in the polymerization of vinyl chloride. J Vinyl Technol 13(1):26–36
159. Allsopp MW (1977) Effect of vinyl chloride injection on the morphology of suspension-polymerized PVC. J Macromol Sci Chem 11(7):1223–1234
160. Cheng J, Langsam MJ (1984) Effect of cellulose suspension agent structure on the particle morphology of PVC. Part II. Interfacial properties. Macromol Sci Chem A21(4):395–409

Adv Polym Sci (2018) 280: 195–214
DOI: 10.1007/12_2017_14
© Springer International Publishing AG 2017
Published online: 25 June 2017

# Trends in Emulsion Polymerization Processes from an Industrial Perspective

Klaus-Dieter Hungenberg and Ekkehard Jahns

**Abstract** We highlight some trends in research on emulsion polymerization, focusing on industrial relevance. The review is restricted to a selected (and somewhat arbitrary) number of topics, namely, the use of renewable raw materials, nanostructured latexes, miniemulsion polymerization, continuous emulsion processes, and recent developments in semibatch processes and removal of volatile organic compounds. Scientific and technical details for many of the topics mentioned here are covered by other articles in this volume.

**Keywords** Continuous polymerization • Emulsion polymerization • Hybrid latices • Miniemulsion polymerization • Multiphase particles • Process control • Renewable resources • VOC removal

## Contents

K.-D. Hungenberg (✉)
Hungenberg Consultant, Ortsstrasse 135, 69488 Birkenau, Germany
e-mail: kdh.consult@yahoo.de

E. Jahns
BASF SE, 67056 Ludwigshafen, Germany
e-mail: ekkehard.jahns@basf.com

# 1 Introduction

Emulsion polymerization is one of the most versatile techniques for the production of synthetic polymers, providing materials of varied chemical composition and application. The final product is known as a dispersion, emulsion, or latex, and all terms are used interchangeably in this article.

The semibatch process, in which monomers and other ingredients (initiator, surfactant, etc.) are metered into the reactor, is now the most commonly used industrial process because it offers the unique opportunity to combine chemically different monomers, such as very hydrophobic and hydrophilic monomers, mono- and multifunctional monomers for crosslinked latexes and films, or monomers of low and high glass transition temperatures. Moreover, this process offers the possibility to generate multiphase polymers of various morphologies (see Fig. 1) and to produce composite and hybrid particles that can incorporate other substances such as pigments, fillers, waxes, and active ingredients.

In emulsion polymerization, particle sizes in the nanometer range are inherent to the process itself. No additional operations such as milling or grinding are needed, or high shear equipment, because polymer chain growth occurs through sequential

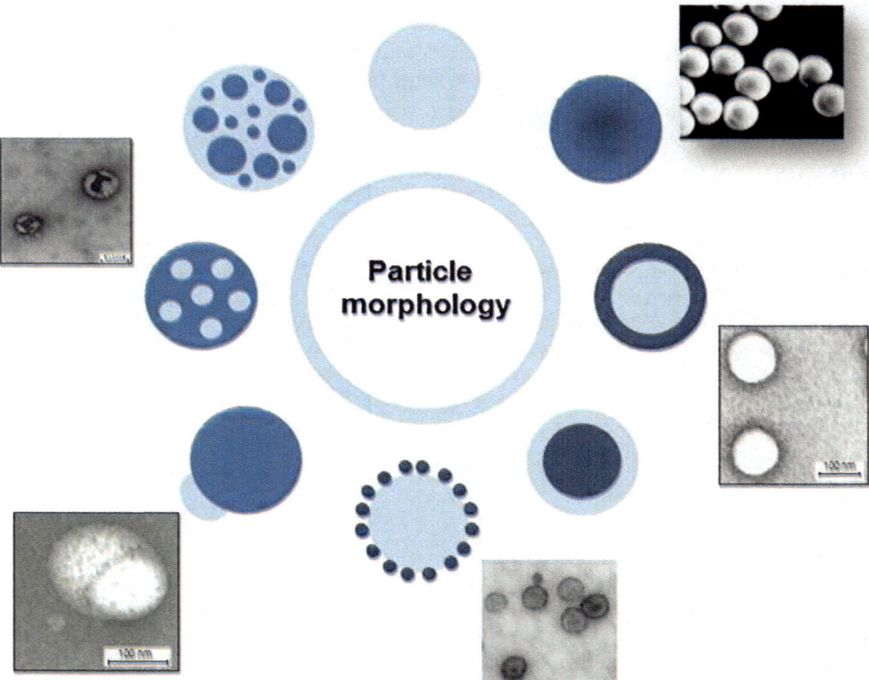

**Fig. 1** Examples of possible morphologies of latex particles

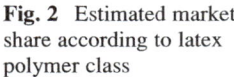

**Fig. 2** Estimated market share according to latex polymer class

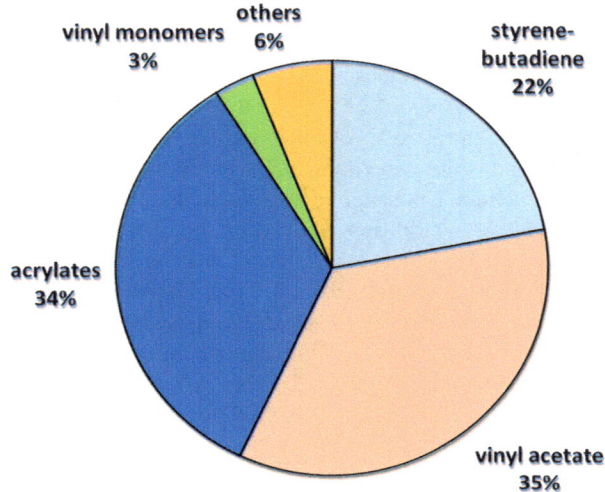

addition of monomers to latex particles at the nanometer scale. The relatively low viscosity of the system, even for solid contents up to approximately 70% in water, enables the application of simple heat removal technologies. Cooling can be accomplished by conduction through the wall, even for large reactors of up to 100 m$^3$, but is often complemented by additional (internal or external) heat exchangers or evaporative cooling. This fairly simple technology results in low investment and operational costs and makes emulsion polymerization technology one of the most efficient polymer processes.

Polymer dispersions are often characterized according to the most abundant monomer(s), as shown in Fig. 2. The numbers represent an estimation of market share using data from various sources. Only those emulsion polymers used as a latex are listed, not polymer types that are isolated from the latex (e.g., by coagulation), examples of which are styrene butadiene rubbers for tires and impact modifiers.

In addition to the main monomers, several comonomers are usually included to improve the properties of the latex; examples include water-soluble monomers such as (meth)acrylic acid and (meth)acrylamide derivatives to improve latex stability. Monomers containing epoxy, hydroxyl, or amine groups incorporate reactive moieties capable of undergoing further reactions (e.g., crosslinking) that are useful in the intended application.

Our daily life would be quite different without the use of latexes. They are applied in all kinds of surface coatings as adhesives and sealants, in fiber bonding for paper and textile applications, as foamed latices for cushioning material, and so on. It is not uncommon for the different classes of latexes shown in Fig. 2 to be used in the same application, demonstrating that the application properties of polymer dispersions are largely determined by their colloidal properties and not just by the polymer backbone, as is the case for example in engineering plastics such as polyamides.

The annual production of synthetic polymer dispersions and latexes in 2014 was estimated to be about 21 million metric (wet) tons per year, and the expected high global growth rate of 3–6% per year (depending on region, application, and source) demonstrates the wide applicability of polymer dispersions.

The vendor landscape is as diverse as the chemistry and properties of polymer dispersions; in addition to leading producers such as BASF, DOW Chemical, Synthomer, Celanese, Wacker, and Arkema, there are hundreds of smaller companies with more specialized portfolios. However, this landscape tends to be volatile and is governed by numerous mergers, joint-ventures, and spin-offs.

Drivers for the increasing use of polymer dispersions and for new product development are manifold and depend on the global region. For example, the increased economic and industrial development in countries such as China and India has increased their use of polymer dispersions, especially in paints, coatings, adhesives, paper, and construction products. Regulatory, environmental, and societal issues are important drivers for the switch from solvent-borne to water-borne polymer systems and there is increasing pressure for use of renewable and biodegradable raw materials.

There is an ongoing trend for polymer dispersions to replace polymer technologies that are less environmentally friendly, such as solvent-based polymers and reactive polymers (e.g., epoxy and polyurethane systems) in a variety of applications. Industrial research is still ongoing to broaden and optimize the applicability of polymer dispersions in these applications, mainly as a result of the wide variety of emulsion polymer chemistries and the flexibility of the process (see for example [1, 2]). Blends and hybrid latexes from a wide range of organic and inorganic materials are beginning to enter the market slowly but surely. Polyurethane acrylic hybrids, epoxy acrylic hybrids, alkyd latex hybrids, and nanosilica acrylic hybrids are just a few examples of products of this class. These new materials combine the chemical and application benefits of these chemistries with the flexibility and environmental benefits of emulsion polymers.

# 2 Renewable Raw Materials

Naturally, there is a trend to replace commonly used monomers derived from fossil feedstock by the same monomers from renewable feedstock [3–5], and there are attempts to introduce new monomers derived from biofeedstock (e.g., linoleic acid [6]). The use of renewable resources involves not only the monomers themselves, but the entire production line. An example is the use of butyl acrylate made from "bio acrylic acid" and "bio butanol" in the production of acrylic emulsions [7].

Moreover, there is a tremendous amount of research into the use of renewable raw materials such as starch and vegetable oils in dispersion formulations [8–11]. However, to date, there has been no real breakthrough because, in many cases, the applications are not yet competitive. Nevertheless, industry is focusing strongly on the use of renewable sources, as shown by the self-defined company strategies of market leaders such as Akzo [12]:

Our strategic ambitions relating to sustainability performance:

- Increase revenue from downstream eco-premium solutions to 20 percent of our revenues by 2020
- Reduce our carbon emissions through the value chain by 25 to 30 percent per ton by 2020 (2012 base)
- Improve resource efficiency across the full value chain

and also from published accomplishments of companies such as BASF [13]:

To manufacture the binders of the Acronal® brand, the company replaces one hundred percent of the fossil resources used at the beginning of the production process with renewable raw materials.

These are just two examples of the many producers of polymer dispersions that are following similar activities.

It is beyond the scope of this article to review all such developments, but it is worth mentioning that a common problem in the use of feedstock from bioresources is the reproducibility of the feedstock. Irregularities in the feedstock can influence not only reaction rates but also the sensitive colloidal properties during polymerization; On-line measurement and control techniques are applied to detect irregularities in the feedstock as early as possible, and to enable appropriate countermeasures to be taken.

# 3   Multiphase and/or Composite Particles

Multiphase and/or composite latexes offer the possibility of combining materials with different properties, which are often contradictory or complementary. Reviews on various classes of hybrid latexes have been published recently, for example, by van Herk and Landfester [14] and Asua [15]. Multiphase particles have been synthesized and used for a broad variety of applications that include typical high-volume applications for latexes such as adhesives, coatings with improved barrier or anticorrosion properties, and pigmented coatings; more specialized applications such as tissue engineering, self-healing coatings, and encapsulation of active components for "medium-sized" products; and low-volume, high-value products for gene and drug delivery.

The presence of multiple phases in latex particles may be the result of different factors. If different (co)polymers are synthesized within one particle, they may phase-separate during or after polymerization. The miscibility of different polymers can be estimated from solubility parameters and sophisticated equation-of-state models. Such coexistence of different polymers forming various phases in one particle can also occur in batch copolymerization if the reactivity ratios and/or solubilities of the monomers are very different, resulting in a dramatic shift in the composition of the polymer during polymerization. However, multiphase polymers are usually produced in a more defined way by stepwise semibatch polymerization, whereby various comonomer mixtures are sequentially fed into the reactor. This procedure does not necessarily mean that a core–shell polymer is produced according to the sequence of monomer addition; a wide variety of different morphologies are possible, even for the same monomer system.

The resulting morphology depends on the interplay between polymerization kinetics and the thermodynamics of the system, and is therefore governed by the various physicochemical properties of the system (miscibility of polymers; solubility of monomers in both water and latex particles; degree of polymerization, grafting, and crosslinking; entry and exit rates of radicals into and out of latex particles; glass transition temperature, etc.) and by the polymerization process itself (relative rates of polymerization, "cross-polymerization," mass transport and swelling of the latex particles, temperature, order of monomer feeds, etc.). Several attempts have been made to describe these systems and the resulting families of particle morphologies is shown in Fig. 1 [16–19].

Latexes with heterogeneous morphologies that combine a soft, elastic phase with a hard phase of high glass transition temperature are, for example, used as impact modifiers or in wood coating binders to improve performance. In the latter application, the requirements for low film-forming temperature and high elasticity together with early particle cohesion and good block resistance of the drying coatings are provided by multistage latexes, which generated early acceptance in the market [20]. Hemispheric and multiglobule latex particles are useful as binders to help with contradictory properties such as adhesion versus block resistance (Fig. 3).

Another method employed to generate hybrid systems is to dissolve a polymer (or pre-polymer) in a monomer mixture and then to polymerize the mixture (e.g., via miniemulsion technology [21]). Such hybrid systems may consist of, for example, alkyds, epoxies, polyesters, or polyurethanes. Polyurethanes have attracted enormous attention in the scientific literature in recent years [22]. Polyurethane/acrylic hybrid dispersions, for example, are becoming increasingly important in coating applications. The polyurethane component is acknowledged for its outstanding elasticity whereas the acrylic component enhances outdoor resistance. Not all hybrids are "real" hybrids, which are synthesized together, but are mixtures of separately produced dispersions. The properties of real hybrids can be outstanding, but the high cost of the urethane raw materials is still slowing market acceptance.

A special case in emulsion polymerization is the incorporation of a monomer-induced swelling step in the synthetic scheme. An example of such a scheme is the polymerization of a crosslinked polystyrene latex, followed by swelling of this crosslinked particle with styrene, and polymerization of this second monomer portion to yield "Janus" particles with a hydrophobic surface on one side and a hydrophilic surface on the other side (Fig. 4).

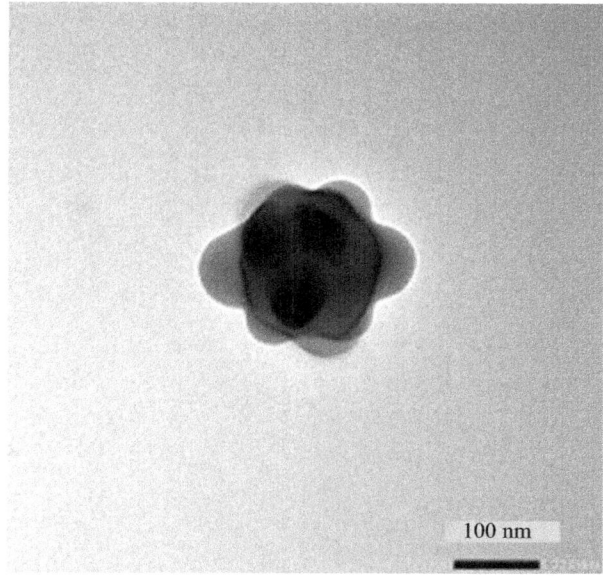

**Fig. 3** A 200 nm "raspberry" latex particle (Source: N. Rajabalinia and J.M. Asua, POLYMAT, San Sebastian)

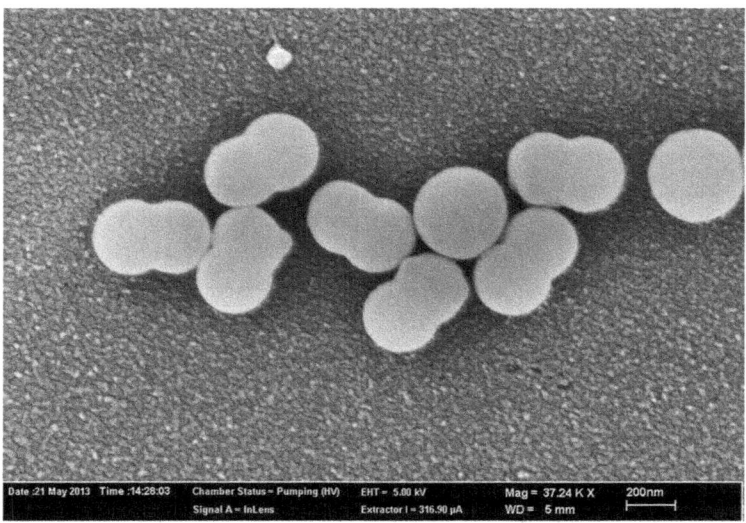

**Fig. 4** Janus particles made by swelling emulsion polymerization (BASF SE)

Another type of structured latex particle encloses inorganic material within the latex. These functional materials are manifold and filler materials range from pigments, silicates, magnetic ferrites, and catalysts to, astonishingly, air.

**Fig. 5** AQACell® DS6299
from BASF SE

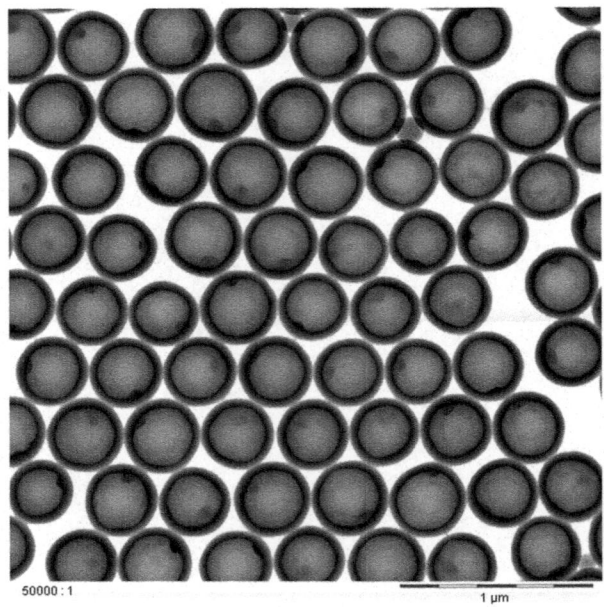

50000 : 1                                                                    1 µm

   Some of the most sophisticated latex polymer products are used in product lines
such as Ropaque® (Dow Chemical) and AQACell® (BASF SE) (see Fig. 5). Hollow
spheres are synthesized using an alkali-swellable, acid-rich core, which is then
encapsulated by a hard polystyrene shell to yield latex particles with a swollen
hydrogel core. The spheres are used in aqueous paint formulations and, as the paint
dries, the hydrogel dries out to give an air void without collapse of the shell. The
very low refractive index of the air void ($n_D \approx 1.00$) against the high refractive
index of the polystyrene shell ($n_D \approx 1.59$) and binder matrix, together with the
optimum particle size of about 350 nm, results in high opacity of the formulation by
intense light scattering, creating a lightweight "plastic white pigment." This is in
contrast to conventional $TiO_2$ white pigments, which scatter light because of the
very high refractive index of $TiO_2$ ($n_D \approx 2.5$–2.9) against the much lower refractive
index of the binder matrix. Combining both technologies, $TiO_2$ and the "air void"
latex together give optimum performance in white paints and coatings.
   However, in spite of these possibilities, there are very few commercial applications for multiphase particles on the market (excluding applications for impact-
modified thermoplastics, wood coating paints, and clear varnishes). One may ask
what the reason is for such a low uptake of this technology, and reach the
conclusion that the application advantages do not balance the higher effort required
to produce these products. Because the most usual method for their production is
via the miniemulsion process, the commercial viability of this process must be
improved.

# 4 Miniemulsion Polymerization

Classical emulsion polymerization is a very versatile polymerization method but there are two drawbacks caused by monomer solubility in water. With highly water-soluble monomers, aqueous-phase polymerization cannot be prevented and, therefore, water-soluble monomers are often polymerized by "inverse emulsion" polymerization, in which water droplets containing the monomer are polymerized in an oil phase. With monomers of low water solubility, the necessary monomer transport from monomer droplets to the locus of polymerization (polymerizing latex particles) is hindered, resulting in a low rate of polymerization. This problem can be overcome in miniemulsion systems by switching the locus of polymerization from the latex particles to the monomer droplets. To achieve this, the monomer droplets must be much smaller (0.05–0.5 µm) than in traditional emulsion polymerization (usually 1–100 µm). Therefore, high shear must be applied to the monomer–water system (e.g., by sonication, rotor-stator equipment, or high pressure homogenizers) in the presence of a suitable combination of surfactant and co-stabilizer. The co-stabilizer is usually a hydrophobic liquid, such as hexadecane, which prevents Ostwald ripening. The small size of the monomer droplets and the resulting high interfacial area has two consequences. First, because of the high surface area, stabilization of the droplet consumes a large portion of surfactant, so less surfactant (ideally none) is left to form micelles, which is (in many cases) where polymerization starts in classical emulsion polymerization. Second, the high surface area of monomer droplets makes them successful in competing with any remaining micelles in scavenging water-borne radicals. Thus, overall, the nucleation mechanism is (ideally completely) switched to droplet nucleation – a mechanism that is usually avoided in classical emulsion polymerization. There are several reviews available on this topic (e.g., Schork et al. [21], van Herk [23], and Asua [15]).

Originally, the advantage of miniemulsions was seen as enabling polymerization of highly hydrophobic monomers in emulsion. This is to some extent still true; for example, the incorporation of long-chain alkyl (meth)acrylates into acrylate formulations can increase their resistance to hydrolysis in acidic environments.

Over time, it has become apparent that the main advantage of miniemulsion polymerization is its ability to produce very complex latex particles, for example, by including various components in the monomer droplets and then polymerizing them, resulting in multiphase/composite particles. However, despite these advantages and widespread research in academia, research institutions, and industry, which has been carried out for more than 40 years since the first publication on this topic [24], the application of miniemulsion polymerization in industry is still rare and limited to some specialty products; examples include UV filters and light stabilizers based on "novel encapsulated additive technology" (NEAT) [25], the Tinuvin® DW family of light stabilizers, and a UV filter for cosmetic application (Tinosorb® S, from BASF SE).

In spite of the importance of miniemulsion polymerization for the synthesis of nanostructured particles and the tremendous interest in this topic in academic research, there are relatively few patents on this subject. There is a striking discrepancy between the scientific and academic interest in miniemulsion polymerization and its commercial importance if one takes the number of scientific publications and the number of patents as the respective measures.

A search for scientific publications on the topic "miniemulsion polymerization," using Google Scholar for the period 2000–2016, yielded about 16,000 articles (excluding patents and citations). A patent search (ESPACENET of the European Patent Office) on this topic gave approximately 190 results in the worldwide database in the title or abstract; moreover, many of the applicants were from academia. For the same period, there were about 10,000 patents on emulsion polymerization and approximately 85,000 scientific articles.

Thus, for emulsion polymerization the ratio of patent applications to scientific publications was about 1:10, whereas for miniemulsion polymerizations it was 1:100.Even though this is a very rough measure or indication, it does show that real-world applications of miniemulsion polymerization are rare, as are the chances of a breakthrough in the near future.

The reasons have been extensively discussed by Asua [15] and will not be repeated here in detail, but we wish to point out the two main drawbacks that must be overcome before miniemulsion polymerization can become a successful and widespread method for the production of structured nanoparticles.

First, there is lack of knowledge concerning details of the structure required to produce structured nanoparticles, which have superior application properties compared with latexes produced by traditional emulsion polymerization. Important structural aspects are the molecular structure (molar mass distribution, composition distribution, branching, crosslinking, etc.), particle size distribution, and particle morphology in particular.

Second, the miniemulsification step, together with control of droplet size distribution and droplet stabilization throughout the whole process, is crucial for success. A technically and commercially competitive/viable miniemulsification can be summarized as follows:

- Emulsification of 50% or more of an organic phase (eventually containing nanosized particles)
- Viscosity ranging from 1 mPa·s for pure monomer systems to around $10^3$ mPa·s for systems with preformed polymer
- Droplet size of 200–500 nm
- Narrow droplet size distribution
- Time scale that does not extend typical cycle times for emulsion polymerization
- Production of several tons per hour

There is no generally applicable method available that fulfils all these requirements. Relatively cheap solutions such as static mixers (with low investment and maintenance costs) generally need hundreds of passes to obtain the required droplet size. Sonication is, in principle, an excellent method because the cavitation effect generates shear within the liquid, but it is difficult to achieve droplet sizes below 1 μm and the technique is generally limited to small quantities. More sophisticated and expensive devices are rotor-stator and high-pressure homogenization systems, which, in spite of the additional high investment, do not generally give satisfactory results. Consequently, existing applications are usually restricted to some special applications that fulfil some of the criteria listed above.

From a mechanistic aspect, it is a challenging task to ensure that nucleation occurs in the monomer droplet and does not arise from homogeneous or micellar nucleation. Generally, this must be considered on a case-by-case basis in terms of chemistry and mode of reactor operation.

In summary, although miniemulsion polymerization is an established and specialized form of emulsion polymerization, there is still a lot of work to do with respect to fundamental understanding of structure–property relationships in order to design superior properties for a range of applications. Furthermore, developments in chemical engineering are needed to overcome process challenges.

## 5  Semibatch Emulsion Polymerization

The semibatch approach is the most widely used process for emulsion polymerization of different grades because it offers the highest flexibility and allows control of the polymerization process, rate of polymerization (and, thus, rate of heat production), and properties of the latex, such as composition, particle size distribution, molecular weight distribution, gel content, and morphology.

According to the way monomer is fed to the reaction, one can distinguish between monomer feed and emulsion monomer feed (also known as pre-emulsion feed). Monomer feed requires a minimum of equipment because there is no need for any emulsification apparatus. Because all of the surfactant is charged at the beginning of the process, particle sizes are usually small and thus prone to coagulation. If transport of monomer into the latex particles is not fast enough, droplet nucleation can occur, resulting in unwanted coarse material. Moreover, the rate of polymerization is often limited by the rate of transport of monomer to the latex particles.

Monomer pre-emulsion feed requires an emulsification device, which in most cases is a stirred tank, but static or dynamic mixers are also used (see, for example, Kostansek [26]). The amount of emulsifier in the reactor changes throughout the reaction and must be carefully balanced to allow sufficient surface coverage of latex particles and monomer droplets to guarantee stabilization but should not exceed the critical micelle concentration (CMC) in order to avoid secondary nucleation.

However, in some cases, secondary nucleation is a desired phenomenon and is used to produce latexes with a bimodal size distribution to increase the solid content of the dispersion. A second generation of particles can be initiated by addition of "seed" particles, or by addition of surfactant to rapidly exceed the CMC and start a second generation by micellar nucleation [27–29].

With respect to the relative rates of polymerization (rP) and monomer feed (rF), one can distinguish between two extreme cases: monomer-flooded, with rP < rF, and monomer-starved, with rP > rF. Although the first mode offers the opportunity to run at the highest polymerization rate as it operates at the highest monomer concentration in the latex particles, it is seldom used. Disadvantages include the formation of polymerized monomer droplets, increased risk of runaway reactions, poor control of copolymer composition, poor particle size control, and reactor fouling.

Using monomer-starved conditions, the process is mass-transport controlled and thus more easily accessible to reaction engineering measures and control actions. The rate of polymerization, and hence the rate of heat production, can easily be controlled by the monomer inlet mass flow, as is also the case for copolymer composition (identical to the feed composition under starved conditions). Semibatch polymerization is the method of choice for the production of structured particles, such as core–shell particles, by emulsion polymerization.

The disadvantage, however, is obvious – the process runs with a lower polymerization rate than would be possible from the kinetic point of view. Moreover, the rate of heat production must always be kept lower than the heat removal capacity of the reactor. In most cases, recipes for emulsion polymerization are developed for constant monomer feed rates and, generally, constant reaction temperature (without taking into account disturbances during processing of an individual batch, and variations from batch to batch). Disturbances include an increase in fill level, increase in viscosity, variation in feed temperature, varying levels of impurities, and fouling of the reactor wall. The operation of a large-scale reactor with such a fixed recipe must be rather conservative to capture all these variations.

Safety measures must be taken to safeguard the reactor against unforeseen disturbances (e.g., failure of pump, cooling water, or stirrer systems). In addition to conventional mechanical devices such as rupture discs or safety valves, an on-line safety concept [30, 31] can be applied, which limits the amount of unreacted monomer in the reactor by monitoring conversion using on-line calorimetry and by quantitatively updating the actual hazard potential in terms of runaway temperature and pressure. If a potential violation is observed, feeds to the reactor are interrupted or slowed.

There are also examples of emulsion and inverse emulsion polymerization that do not involve heat removal and the polymerization is performed in an inherently safe adiabatic and very fast process [32, 33]. A minimum of safety measures are necessary, but overall controllability is poor.

On-line conversion information, obtained by on-line calorimetry or other methods (see for example Frauendorfer et al. [34] and Fonseca et al. [35]), is a valuable source of information on the state of the process and a prerequisite for improving the performance of a reactor.

Figure 6 gives an example of process variations during one batch emulsion polymerization and between different batches of the same product in terms of the cooling temperature required to keep the reactor temperature at the desired value. During the first half of the reaction, the available cooling capacity is not fully exploited; in principle, one could feed monomer faster and lower the cooling temperature appropriately. Moreover, different batches behave differently, for example because of different states of fouling or different feed temperatures, and some could be run faster than others.

**Fig. 6** Cooling water temperature during the course of an emulsion polymerization with fixed feed rates and constant reactor temperature for various repeat batches

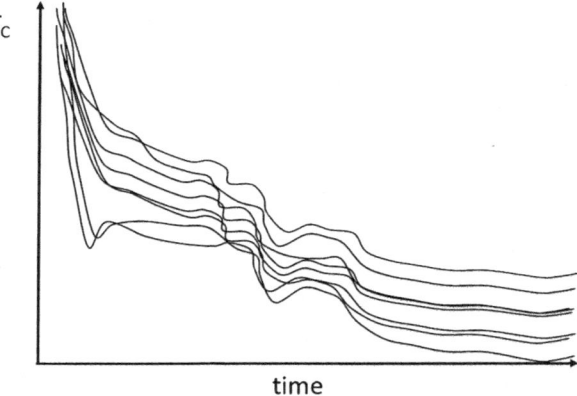

$T_C$

time

To optimize the yield in such a situation, simultaneous control of monomer feed rates and cooling water temperature is necessary. Pelz et al. [36] describe a control scheme for time-optimal operation of a semibatch emulsion polymerization reactor, taking into account jacket temperature constraints and avoidance of a monomer droplet phase. Vicente et al. [37] experimentally verified the on-line control of composition and molar mass during emulsion polymerization in a laboratory-scale reactor. More examples of the application of nonlinear model predictive control (NMPC) to the optimization of (semi)batch processes can also be found in Bonvin et al. [38] and the abovementioned publications.

Despite the highly sophisticated academic research and the promising results in this field, very little has been reported on the industrial application of NMPC for polymer processes in general, and emulsion polymerization in particular. Finkler et al. [39] describe an application for minimization of batch duration in an industrial reactor for solution polymerization, and Graichen et al. [40] report implementation of feedforward control for a polymerization process.

There could be several reasons why NMPC control schemes in (semi)batch polymer processes in industry are not being implemented, and some improvements and changes are required to foster their application. The cost and effort needed for model development, implementation, and maintenance is currently high compared with the benefit, although a 10% reduction in batch time has been reported [36]. Acceptance by plant personnel is important, because the controller actions of an NMPC system might be quite different to those of a conventional proportional-integral-derivative (PID) controller to which the operators are accustomed. The conventional time-controlled recipe (i.e., carrying out the same operation at the same time for every batch) is often part of the quality concept. This concept is violated by NMPC control actions. For example, instead of constant monomer feed rates over a predetermined time interval for every batch of a certain grade, the monomer feed rate is (more or less steadily) changed according to the actual state of the batch (see Fig. 6). Time-optimal NMPC implementations often neglect the properties of the polymer as end-point constraints (e.g., molar mass averages or distributions, composition, branching, or particle size) or, if they do, it is usually done in an open-loop way based on more or less exact models (e.g., Vicente et al. [37] and Gomes [41]), which are prone to uncertainties. Moreover, taking polymer properties as sole end-point constraints is usually not sufficient. Depending on the controller actions, concentrations are prone to changes throughout the process and the accumulated polymer properties may differ accordingly; therefore, the polymer properties should also be set as path constraints in many cases. Moreover, reliable, fast analytical on-line methods for monitoring polymer properties to close the control loop are rare, especially for emulsion polymers where the polymer properties are hidden in the latex particle. There is a strong need for further development of on-line sensors (hard or soft) for monitoring polymer properties and, thus, opening the way to closed control loops.

The overall productivity of the reactor and product quality are determined not only by the efficiency of heat removal from an emulsion polymerization reactor, but also by factors (e.g., homogeneity of the reactor content, coagulum formation) that depend on the fluid dynamic behavior of the reacting mixture. Coupling of population balances with computational fluid dynamics tools [42] to describe the evolution of particle size distribution and, simultaneously, mixing behavior is a valuable method for scale-up and optimization of reactor geometry (stirrer, baffles) and operation conditions such as stirrer speed or position of feeds and their composition.

## 6 Continuous Emulsion Polymerization

There are several continuous emulsion polymerization processes in use, especially for large-scale products such as rubbers [43] and polyvinylchloride, but also for copolymers from vinyl acetate and ethylene (e.g., de Castro and Adams [44], and Hain et al. [45]). In most cases, a series of continuous stirred tank reactors (CSTRs) are used, or combinations of CSTR and tubular reactors. Sometimes, the CSTR is mimicked by a loop tubular reactor [46]. Continuous emulsion processes in reactors with the residence time distribution of a CSTR have to face a serious problem, namely the occurrence of damped or sustained oscillations, which result in multiple steady states for conversion and variations in the properties of the latex and polymer (e.g., particle size or molecular weight and their distributions) [47]. A method of choice to avoid such oscillations is to shift the nucleation process outside the CSTR, either by using a seed latex in the feed or by placing a tubular reactor before the CSTR [48].

In recent years there has been a strong trend to process intensification for polymer production by replacing the conventional and flexible (semi)batch reactors by continuous reactors. Durand and Engell [49] describe the incentives for such a switchover: "In continuous production, on the other hand, the heat removal capacity is larger so that higher reaction rates (solids contents) can be realized, and hence the space-time yield is higher. Moreover, continuous processes can be more tightly controlled and require less cleaning as long as there are no product changeovers. The switch from batch to continuous production is one facet of process intensification which can lead to savings of energy and material, more compact and thus cheaper plants and improved sustainability and better economic performance." A large European research project, "Flexible, Fast, and Future Production Processes (F3 Factory)," led by the European chemical and pharmaceutical companies Bayer, BASF, Arkema, AstraZeneca, Rhodia (now Solvay), and Evonik has been instigated that targets the development of continuous standardized plants of small to medium scale .

Over the years, several promising results on the transfer from batch to continuous processes have been published. The group of Moritz reported continuous emulsion polymerization in reactors with superimposed secondary flow [50]. Rossow et al. [51] describe an example of a tubular reactor. An overview on recent developments in this field is given by Asua [52], together with examples of the transfer from semibatch reactors to a cascade of CSTRs. Other examples relate to the use of sieve-plate columns as a tubular reactor [53].

However, despite these and other successful examples at the laboratory scale, a real breakthrough has not yet been achieved and there are still several economic and technical questions, as well as problems, that need to be resolved.

The fact that polymers, especially emulsion latexes, are products by process is usually mentioned but the consequences are often underestimated. Polymer structure, colloidal properties, and application-related properties are different for different processes. Many latexes are produced as a range of specialized grades, and so an entire product portfolio has to be developed for a new process. Moreover, new research infrastructure, at both laboratory and pilot scales, has to be established for a new process. It must be carefully analyzed whether the product(s) under consideration for a switch to continuous process manufacture are suitable for such a transition in terms of product volume, number of necessary grade changes, stockholding, and many other aspects.

The widespread argument that the use of continuous reactors leads to lower operational and capital costs is not always true and must be carefully considered in each case. The same holds true for the argument that continuous processes yield less out of specification (off-spec) material than semibatch reactors. The use of modern control and automation tools also makes the generation of off-spec material a rare event in batch processes.

There are also a number of technical problems to be resolved, which are more or less problematic depending on the reactor set-up and its residence time distribution and geometry. Problems such as fouling, plugging, and cleaning of the reactor are of operational relevance, whereas conversion, solid content, performance of grade changes, and residual monomer are of economic relevance. The achievable polymer structure (molar mass, composition, branching, crosslinking, and their distributions) and colloidal properties are issues related to product quality.

The problems and questions discussed above for the change from a semibatch to continuous process seem to be especially challenging for emulsion polymerization, with its multiple phases, compared with solution or bulk processes.

# 7  Removal of Volatile Organic Compounds

Regulatory and societal requirements for reduced VOCs in paints, coatings, etc. makes the post-treatment of latexes an important process step. Despite its importance, there are only a few scientific articles on this topic. Araujo et al. [54] review monomer reduction in polymer systems in general. There are, in principle, two ways to reduce VOCs – chemical and physical. Chemical measures usually comprise post-polymerization in the reaction vessel itself or in a separate vessel by adding a fast-reacting initiator system comprising peroxides and reducing agents (so-called redox systems). Ilundain et al. [55] present an experimental and theoretical investigation on the post-polymerization of vinyl acetate dispersions.

At first glance, chemical VOC removal seems to be the method of choice because it can be carried out easily, without any extra investment, in the polymerization reactor itself. Alternatively, removal can take place in a post-reactor, which can be of a much simpler and cheaper design than the polymerization reactor itself because it operates at ambient pressure and without heat removal. However, post-polymerization can only reduce the concentration of polymerizable monomers; the concentration of other VOCs (e.g., saturated analogues of the monomers) remains unchanged. Furthermore, low molecular weight oligomers/polymers resulting from the high flux of radicals during chemical VOC removal may change the properties of the latex. Radicals from the post-polymerization initiator system might not (or to a limited extent) enter the latex particles, so only water-soluble monomers are significantly reduced and hydrophobic monomers remain mostly unaffected. Decomposition products from organic peroxides and/or reducing agents contribute to the overall VOCs. As the initiator system reacts rapidly, even before being homogeneously distributed in the reactor, locally increased levels of electrolytes or increased pH values can cause coagulum formation. This inhomogeneity increases the use of the initiator well above the theoretical requirement.

Consequently, in many cases, physical measures are used alone, or together with post-polymerization techniques to reduce VOCs in polymer dispersions. Kechagia et al. [56] present a study on the combination of both methods. Stripping of aqueous dispersions, usually with the aid of steam, air, or nitrogen, is the most frequently used physical method. Different types of apparatus can be employed and vessels [57], columns [58], and thin-layer evaporators [59] have been described (see also Englund [60]). Using this technology, nonpolymerizable VOCs can also be removed, but this process step requires additional investment and still has an obvious drawback – reduction of high-boiling compounds is limited because of their high boiling points and poor water solubility. Thus, it is difficult to remove compounds such as long alkyl chain (meth)acrylic esters, or 4-phenyl cyclohexene and 4-vinyl cyclohexene, which are Diels–Alder reaction products from styrene and butadiene (in styrene–butadiene dispersions).

In addition to steam and air, supercritical carbon dioxide has been used as processing aid in the devolatilization of latexes [61, 62]. However, the study was limited to the removal of monomers such as styrene and methyl methacrylate, and did not investigate the removal of high-boiling components.

# 8  Summary

Even though the emulsion polymerization process is over 100 years old, there is still a lack of detailed knowledge and understanding of the process, which makes it a challenge from both the scientific and industrial points of view. The transfer from scientific insights and findings in academia to an industrially viable application is not straightforward and usually takes time and the serendipitous occurrence of suitable opportunities. Such opportunities may come from customer demands, legislative regulations, societal requests, raw material issues, market pressure, and many more. Last, but not least, the implementation of new technologies must have an economic benefit for industry.

# References

1. Mckenna TF, Charleux B, Bourgeat-Lami E, D'agosto F, Lansalot M (2014) Novel technologies and chemistries for waterborne coatings. J Coat Technol Res 11:131–141
2. Overbeek A (2009) Polymer heterogeneity in waterborne coatings. J Coat Technol Res 7 (1):1–21
3. Chung H, Yang JE, Ha JY, Chae TU, Shin JH, Gustavsson M, Lee SY (2015) Bio-based production of monomers and polymers by metabolically engineered microorganisms. Curr Opin Biotechnol 36:73–84
4. Dubé MA, Salehpour S (2014) Applying the principles of green chemistry to polymer production technology. Macromol React Eng 8:7–28
5. Gandini A, Lacerda TM (2015) From monomers to polymers from renewable resources: recent advances. Prog Polym Sci 48:1–39
6. Roberge S, Dubé MA (2016) Infrared process monitoring of conjugated linoleic acid/styrene/butyl acrylate bulk and emulsion terpolymerization. J Appl Polym Sci 133:n/a–n/a
7. Global Markets Insights (2017) Bio acrylic acid market size. Report GMI341. Global Market Insights, Delaware. https://www.gminsights.com/industry-analysis/bio-acrylic-acid-market. Accessed 24 Apr 2017
8. Bao Z, Li W, Fu Z, Chen L (2016) A review of the application of renewable resources in preparing acrylic polymer latex. Polym Renewable Resour 7:13
9. Derksen JTP, Petrus Cuperus F, Kolster P (1996) Renewable resources in coatings technology: a review. Prog Org Coat 27:45–53
10. Lligadas G, Ronda JC, Galià M, Cádiz V (2013) Renewable polymeric materials from vegetable oils: a perspective. Mater Today 16:337–343

11. Vähä-Nissi M, Laine C, Talja R, Mikkonen H, Hyvärinen S, Harlin, A (2011) Aqueous dispersions from biodegradable/renewable polymers. http://www.tappi231.tappi.org/content/events/10PLACE/Aqueous.pdf. Accessed 24 Apr 2017
12. Akzo Nobel (2017) Our strategy. Akzo Nobel, Arnhem. https://www.akzonobel.com/about-us/strategy. Accessed 24 Apr 2017
13. Schnorbus P (2016) From biomass to dispersions: For the first time BASF produces binders for interior paints based on the mass balance process. News release, 17 March 2016. BASF, Ludwigshaven. https://www.basf.com/documents/corp/en/news-and-media/news-releases/2016/03/P155e_Dispersions_Mass_Balance_Process.pdf. Accessed 24 Apr 2017
14. Van Herk AM, Landfester K (2010) Hybrid latex particles: preparation with (mini) emulsion polymerization. Springer, Berlin
15. Asua JM (2014) Challenges for industrialization of miniemulsion polymerization. Prog Polym Sci 39:1797–1826
16. Akhmatskaya E, Asua JM (2012) Dynamic modeling of the morphology of latex particles with in situ formation of graft copolymer. J Polym Sci A Polym Chem 50:1383–1393
17. Hamzehlou S, Leiza JR, Asua JM (2016) A new approach for mathematical modeling of the dynamic development of particle morphology. Chem Eng J 304:655–666
18. Karlsson OJ, Stubbs JM, Carrier RH, Sundberg DC (2003) Dynamic modeling of non-equilibrium latex particle morphology development during seeded emulsion polymerization. Polym React Eng 11:589–625
19. Sundberg DC, Durant YG (2003) Latex particle morphology, fundamental aspects: a review. Polym React Eng 11:379–432
20. Procopio L, Vielhauer L, Greyson E, Hejl A (2012) Designing latex film morphology to optimize wood coating performance. JCT CoatingsTech 9:32–41
21. Schork FJ, Luo Y, Smulders W, Russum JP, Butté A, Fontenot K (2005) Miniemulsion polymerization. In: Okubo M (ed) Polymer particles. Springer, Berlin, Heidelberg
22. Lopez A, Degrandi-Contraires E, Canetta E, Creton C, Keddie JL, Asua JM (2011) Waterborne polyurethane–acrylic hybrid nanoparticles by miniemulsion polymerization: applications in pressure-sensitive adhesives. Langmuir 27:3878–3888
23. Van Herk AM (2010) Historical overview of (mini)emulsion polymerizations and preparation of hybrid latex particles. In: Van Herk AM, Landfester K (eds) Hybrid latex particles: preparation with (mini)emulsion polymerization. Springer, Berlin, Heidelberg
24. Ugelstad J, El-Aasser MS, Vanderhoff JW (1973) Emulsion polymerization. Initiation of polymerization in monomer droplets. J Polym Sci Polym Lett Ed 11:503–513
25. Schaller C, Rogez D, Braig A (2012) Organic vs inorganic light stabilizers for waterborne clear coats: a fair comparison. J Coat Technol Res 9:433–441
26. Kostansek E (2003) Emulsions. In: Kirk-Othmer encyclopedia of chemical technology. Wiley, Hoboken
27. Ai Z, Deng R, Zhou Q, Liao S, Zhang H (2010) High solid content latex: preparation methods and application. Adv Colloid Interf Sci 159:45–59
28. Boutti S, Graillat C, Mckenna TF (2005) High solids content emulsion polymerisation without intermediate seeds. Part III. Reproducibility and influence of process conditions. Polymer 46:1223–1234
29. Guyot A, Chu F, Schneider M, Graillat C, Mckenna T (2002) High solid content latexes. Prog Polym Sci 27(8):1573–1615
30. Kröner H, Klostermann R, Birk J, Hauff T (2002) Method for continuously monitoring and controlling the monomer conversion during emulsion polymerization. US6498219
31. Manders LG, Meister M, Bausa J, Hungenberg K-D (2011) Online model-based process safety concepts in polymerization. Macromol Symp 302:289–296
32. Gao J, Kong XM, Hungenberg KD, Schmidt-Thummes J (2007) Method for producing aqueous polymer dispersions. US10566248
33. Kane J, Durham JF (2016) Continuous adiabatic inverse emulsion polymerization process. US9434793

34. Frauendorfer E, Wolf A, Hergeth WD (2010) Polymerization online monitoring. Chem Eng Technol 33:1767–1778
35. Fonseca GE, Dubé MA, Penlidis A (2009) A critical overview of sensors for monitoring polymerizations. Macromol React Eng 3:327–373
36. Pelz K, Brandt H, Finkler TF, Engell S (2012) Scheme for time-optimal operation of semi-batch emulsion polymerization reactors. IFAC Proc Vol 45:239–244
37. Vicente M, Leiza JR, Asua JM (2003) Maximizing production and polymer quality (MWD and composition) in emulsion polymerization reactors with limited capacity of heat removal. Chem Eng Sci 58:215–222
38. Bonvin D, Srinivasan B, Hunkeler D (2006) Control and optimization of batch processes. IEEE Control Syst 26:34–45
39. Finkler TF, Kawohl M, Piechottka U, Engell S (2012) Realization of online optimizing control in an industrial polymerization reactor. IFAC Proc Vol 45:11–18
40. Graichen K, Hagenmeyer V, Zeitz M (2006) Feedforward control with online parameter estimation applied to the Chylla–Haase reactor benchmark. J Process Control 16:733–745
41. Gomes VG (2010) Advanced monitoring and control of multi-monomer system in emulsion polymerization. Macromol React Eng 4:672–681
42. Pohn J, Cunningham M, Mckenna TFL (2013) Scale-up of emulsion polymerization reactors part II – simulations and interpretations. Macromol React Eng 7:393–408
43. Obrecht W, Lambert J-P, Happ M, Oppenheimer-Stix C, Dunn J, Krüger R (2011) Rubber, 4. Emulsion rubbers. In: Ullmann's encyclopedia of industrial chemistry. Wiley-VCH, Weinheim
44. De Castro LBR, Adams D (2015) Vinyl ester/ethylene copolymer dispersions prepared by continuous tubular emulsion polymerization for coating carpet products. US14399742
45. Hain J, Kotschi U, Weitzel HP (2013) Process for continuous emulsion polymerization. US13505921
46. De Castro LBR, Adams D (2013) Vinyl ester/ethylene copolymer dispersions prepared by continuous tubular emulsion polymerization for coating carpet products. Patent WO2013IB01471 20130515
47. Rawlings JB, Ray WH ((1987) Emulsion polymerization reactor stability: simplified model analysis. AICHE J 33:1663–1677
48. Lee HC, Poehlein GW (1986) Continuous tube-CSTR reactor system for emulsion polymerization kinetic studies. Chem Eng Sci 41:1023–1030
49. Durand A, Engell S (2016)) Batch to Conti transfer of polymer production processes. Macromol React Eng 10:308–310
50. Pauer W, Moritz H-U (2006) Continuous reactor concepts with superimposed secondary flow – polymerization process intensification. Macromol Symp 243:299–308
51. Rossow K, Bröge P, Lüth FG, Joy P, Mhamdi A, Mitsos A, Moritz H-U, Pauer W (2016) Transfer of emulsion polymerization of styrene and n-butyl acrylate from semi-batch to a continuous tubular reactor. Macromol React Eng 10:324–338
52. Asua JM (2016) Challenges and opportunities in continuous production of emulsion polymers: a review. Macromol React Eng 10:311–323
53. Scholtens CA, Meuldijk J, Drinkenburg AAH (2001) Production of copolymers with a predefined intermolecular chemical composition distribution by emulsion polymerisation in a continuously operated reactor. Chem Eng Sci 56:955–962
54. Araujo PHH, Sayer C, Poco JGR, Giudici R (2002) Techniques for reducing residual monomer content in polymers: a review. Polym Eng Sci 42:1442–1468
55. Ilundain P, Da Cunha L, Salazar R, Alvarez D, Barandiaran MJ, Asua JM (2002) Postpolymerization of vinyl acetate-containing latexes. J Appl Polym Sci 83:923–928
56. Kechagia Z, Kammona O, Pladis P, Alexopoulos AH, Kiparissides C (2011) A kinetic investigation of removal of residual monomers from polymer latexes via post-polymerization and nitrogen stripping methods. Macromol React Eng 5:479–489
57. Humme G, Plato H, Ott KH, Kowitz F, Hagenberg, P (1983) Process for the removal of residual monomers from ABS polymers. US4399273A

58. Heider W, Huebinger WD, Keller PD, Eiden UD (1998) Counterflow column, process plant and process for reducing volatiles in dispersions. DE19716373A1
59. Wilhelm G, Hurm K, Hirsch RB, Jaschke A, Marzolph H (1976) Process for separating and recovering residual monomers from aqueous suspensions of acrylonitrile polymers. US3980529
60. Englund SM (1981) Monomer removal from latex. Chem Eng Prog 77:55–59
61. Aerts M (2012) Residual monomer reduction in polymer latex products by extraction with supercritical carbon dioxide. TU Eindhoven, Eindhoven
62. Kemmere MF, Meyer T (2006) Supercritical carbon dioxide: in polymer reaction engineering. Wiley, Hoboken

# Index

215

Printed by Printforce, the Netherlands